Communications
in Computer and Information Science   541

Commenced Publication in 2007
Founding and Former Series Editors:
Alfredo Cuzzocrea, Dominik Ślęzak, and Xiaokang Yang

More information about this series at http://www.springer.com/series/7899

Fernando Koch · Christian Guttmann
Didac Busquets (Eds.)

Advances in Social Computing and Multiagent Systems

6th International Workshop
on Collaborative Agents Research and Development, CARE 2015
and Second International Workshop
on Multiagent Foundations of Social Computing, MFSC 2015
Istanbul, Turkey, May 4, 2015
Revised Selected Papers

Springer

Editors
Fernando Koch
Samsung Research Institute
Campinas
Brazil

Didac Busquets
Transport Systems Catapult
Milton Keynes
UK

Christian Guttmann
UNSW
Sydney
Australia

and

Karolinska Institute
Stockholm
Sweden

ISSN 1865-0929   ISSN 1865-0937 (electronic)
Communications in Computer and Information Science
ISBN 978-3-319-24803-5 ISBN 978-3-319-24804-2 (eBook)
DOI 10.1007/978-3-319-24804-2

Library of Congress Control Number: 2015950868

Springer Cham Heidelberg New York Dordrecht London

Printed on acid-free paper

Springer International Publishing AG Switzerland is part of Springer Science+Business Media
(www.springer.com)

Preface

This volume comprises the joint proceedings of two workshops that were hosted in conjunction with the International Conference on Autonomous Agents and Multiagent Systems (AAMAS 2015)[1]: the 6th International Workshop on Collaborative Agents Research and Development (CARE 2015)[2] and the Second International Workshop on Multiagent Foundations of Social Computing (MFSC 2015)[3]. The events took place on May 4, 2015, in Istanbul, Turkey.

Both events promoted discussions around the state-of-the-art research and application of multiagent system technology. CARE and MFSC addressed issues in relevant areas of social computing such as smart societies, social applications, urban intelligence, intelligent mobile services, models of teamwork and collaboration, as well as many other related areas. The workshops received contributions ranging from top-down experimental approaches and a bottom-up evolution of formal models and computational methods. The research and development discussed is a basis of innovative technologies that allow for intelligent applications, collaborative services, and methods to better understand societal interactions and challenges.

The theme of the "CARE for Social Apps and Ubiquitous Computing" workshop focused on computational models of social computing. Social apps aim to promote social connectedness, user friendliness through natural interfaces, contextualization, personalization, and "invisible computing." A key question was on how to construct agent-based models that better perform in a given environment. The discussion revolved around the application of agent technology to promote the next generation of social apps and ubiquitous computing, with scenarios related to ambient intelligence, urban intelligence, classification and regulation of social behavior, and collaborative tasks.

The "Multiagent Foundations of Social Computing" workshop focused on multiagent approaches around the conceptual understanding of social computing, e.g., relating to its conceptual bases, information and abstractions, design principles, and platforms. The discussion was around models of social interaction, collective agency, argumentation information models and data analytics for social computing, and related areas.

The workshops promoted international discussion forums with submissions from different regions and Program Committee members from many counters in Europe (The Netherlands, Greece, France, Luxembourg, Sweden, Spain, UK, Ireland, Italy, Portugal), Asia (Turkey, Singapore), Oceania (Australia, New Zealand), and the Americas (Brazil, Colombia, USA). The CARE 2015 workshop received 14 papers submitted through the workshop website from which we selected five papers for publication, all

---

[1] http://www.aamas2015.com/
[2] http://www.care-workshops.org/
[3] http://www.lancaster.ac.uk/staff/chopraak/mfsc-2015/

being republished as extended versions in this volume. MFSC 2015 selected seven papers for publication, all being promoted as extended versions.

The papers selected for this volume are representative research projects around the aforementioned methods. The selections highlight the innovation and contribution to the state of the art, suggesting solutions to real-world problems as applications built on the proposed technology.

In the first paper, "Automated Negotiation for Traffic Regulation," Garciarz et al. propose a mechanism based on coordination to regulate traffic at an intersection. This approach is distributed and based on automated negotiation. Such technology would allow us to replace classic traffic-light intersections in order to perform a more efficient regulation by taking into account various kinds of information related to traffic or vehicles, and by encouraging cooperation.

The second paper, "Towards a Middleware for Context-Aware Health Monitoring," by Oliveira et al., introduces a new model to correlate mobile sensor data, health parameters, and situational and/or social environment. The model works by combining environmental monitoring, personal data collecting, and predictive analytics. The paper presents a middleware called "Device Nimbus" that provides the structures with which to integrate data from sensors in existing mobile computing technology. Moreover, it includes the algorithms for context inference and recommendation support. This development leads to innovative solutions in continuous health monitoring, based on recommendations contextualized in the situation and social environment.

The third paper, "The Influence of Users' Personality on the Perception of Intelligent Virtual Agents Personality and the Trust Within a Collaborative Context," by Hanna and Richards, explores how personality and trust influence collaboration between humans and human-like intelligent virtual agents (IVAs). The potential use of IVAs as team members, mentors, or assistants in a wide range of training, motivation, and support situations relies on understanding the nature and factors that influence human–IVA collaboration. The paper presents an empirical study that investigated whether human users can perceive the intended personality of an IVA through verbal and/or non-verbal communication, on one hand, and the influence of the users' own personality on their perception, on the other hand.

The fourth paper, "The Effects of Temperament and Team Formation Mechanism on Collaborative Learning of Knowledge and Skill in Short-Term Projects," by Farhangian et al., introduces a multi-agent model and tool that simulates team behavior in virtual learning environments. The paper describes the design and implementation of a simulation model that incorporates personality temperaments of learners and also has a focus on the distinction between knowledge learning and skill learning, which is not included in existing models of collaborative learning. This model can be significant in helping managers, researchers, and teachers to investigate the effect of group formation on collaborative learning and team performance. Simulations built upon this model allow researchers to gain better insights into the impact of an individual learner's attributes on team performance.

The fifth paper, "Exploring Smart Environments Through Human Computation for Enhancing Blind," by Paredes et al., presents a method for the orchestration of wearable sensors with human computation to provide map metadata for blind navigation. The research has been motivated by the need for innovation toward navigation aids for the blind, which must provide accurate information about the environment and select the best path to reach a chosen destination. The dynamism of smart cities promotes constant change and therefore a potentially dangerous territory for these users. The paper proposes a modular architecture that interacts with environmental sensors to gather information and process the acquired data with advanced algorithms empowered by human computation. The gathered metadata enables the creation of "happy maps" to provide orientation to blind users.

In the sixth paper, "Incorporating Mitigating Circumstances into Reputation Assessment," Miles and Griffiths present a reputation assessment method based on querying detailed records of service provision, using patterns that describe the circumstances to determine the relevance of past interactions. Employing a standard provenance model for describing these circumstances, it gives a practical means for agents to model, record, and query the past. The paper introduces a provenance-based approach, with accompanying architecture, to reputation assessment informed by rich information on past service provision; query pattern definitions that characterize common mitigating circumstances; and an extension of an existing reputation assessment algorithm that takes account of this richer information.

In the seventh paper, "Agent Protocols for Social Computation," Rovatsos et al. propose a data-driven method for defining and deploying agent interaction protocols that is based on using the standard architecture of the World Wide Web. The paper is motivated by the fact that social computation systems involve interaction mechanisms that closely resemble well-known models of agent coordination; current applications in this area make little or no use of agent-based systems. The proposal contributes with message-passing mechanisms and agent platforms, thereby facilitating the use of agent coordination principles in standard Web-based applications. The paper describes a prototypical implementation of the architecture and experimental results that prove it can deliver the scalability and robustness required of modern social computation applications while maintaining the expressiveness and versatility of agent interaction protocols.

The eighth paper, "Negotiating Privacy Constraints in Online Social Networks," by Mester et al., proposes an agreement platform for privacy protection in Online Social Networks where privacy violations that take place result in users' concern. The research proposes a multiagent-based approach where an agent represents a user. Each agent keeps track of its user's preferences semantically and reasons on privacy concerns effectively. The proposed platform provides the mechanisms with which to automatically settle differences in the privacy expectations of the users.

The ninth paper, "Agent-Based Modeling of Resource Allocation in Software Projects Based on Personality and Skill," by Farhangian et al., presents a simulation model for assigning people to a set of given tasks. This model incorporates the personality and skill of employees in conjunction with the task attributes such as their dynamism level. The research seeks a comprehensive model that covers all the factors that are involved in the task allocation systems such as teamwork factors and the environment. The proposal aims to provide insights for managers and researchers, to investigate the effectiveness of (a) selected task allocation strategies and (b) of employees and tasks with different attributes when the environment and task requirements are dynamic.

In the tenth paper, "On Formalizing Opportunism Based on Situation Calculus," Lou et al. propose formal models of opportunism, which consist of the properties knowledge asymmetry, value opposition, and intention, based on situation calculus in different context settings. The research aims to formalize opportunism in order to better understand the elements in the definition and how they constitute this social behavior. The proposed models can be applied to the investigation of on behaviour emergence and constraint mechanism, rendering this study relevant for research around multiagent simulation.

In the next paper, "Programming JADE and Jason Agents Based on Social Relationships Using a Uniform Approach," Baldoni et al. propose to explicitly represent agent coordination patterns in terms of normatively defined social relationships, and to ground this normative characterization on commitments and on commitment-based interaction protocols. The proposal is put into effect by the 2COMM framework. Adapters were developed for allowing the use of 2COMM with the JADE and the JaCaMo platforms. The paper describes how agents can be implemented in both platforms by relying on a common programming schema, despite them being implemented in Java and in the declarative agent language Jason, respectively.

Finally, the paper "The Emergence of Norms via Contextual Agreements in Open Societies," by Vouros, proposes two social, distributed reinforcement learning methods for agents to compute society-wide agreed conventions concerning the use of common resources to perform joint tasks. The computation of conventions is done via reaching agreements in agents' social context, via interactions with acquaintances playing their roles. The formulated methods support agents to play multiple roles simultaneously; even roles with incompatible requirements and different preferences on the use of resources. The work considers open agent societies where agents do not share common representations of the world. This necessitates the computation of semantic agreements (i.e., agreements on the meaning of terms representing resources), which is addressed by the computation of emergent conventions in an intertwined manner. Experimental results show the efficiency of both social learning methods, even if all agents in the society are required to reach agreements, despite the complexity of the problem scenario.

We would like to thank all the volunteers who made the workshops possible by helping in the organization and in peer reviewing the submissions.

August 2015   Fernando Koch
                                                  Christian Guttmann
                                                    Didac Busquets

Organization

CARE 2015

Organizing Committee

Fernando Koch	Samsung Research Institute, Brazil
Christian Guttmann	UNSW, Australia; Karolinska Institute, Sweden

Program Committee

Amal El Fallah Seghrouchni	University of Pierre and Marie Curie LIP6, France
Andrew Koster	Samsung Research Institute, Brazil
Artur Freitas	PUC-RS, Brazil
Carlos Cardonha	IBM Research, Brazil
Carlos Rolim	Federal University of Rio Grande do Sul, Brazil
Cristiano Maciel	Federal University of Mato Grosso, Brazil
Eduardo Oliveira	The University of Melbourne, Australia
Felipe Meneguzzi	PUC-RS, Brazil
Gabriel De Oliveira Ramos	Federal University of Rio Grande do Sul, Brazil
Gaku Yamamoto	IBM Software Group, USA
Ingo J. Timm	University of Trier, Germany
Jose Viterbo	UFF, Brazil
Kent C.B. Steer	IBM Research, Australia
Liz Sonenberg	The University of Melbourne, Australia
Luis Oliva Technical	University of Catalonia, Spain
Priscilla Avegliano	IBM Research, Brazil
Takao Terano	Tokyo Institute of Technology, Japan
Tiago Primo	Samsung Research Institute, Brazil
Yeunbae Kim	Samsung Research Institute, Brazil

MFSC 2015

Organizing Committee

Amit K. Chopra	Lancaster University, UK
Harko Verhagen	Stockholm University, Sweden
Didac Busquets	Imperial College London, UK

Program Committee

Aditya Ghose	University of Wollongong, Australia
Alexander Artikis	NCSR Demokritos, Greece

Cristina Baroglio	University of Turin, Italy
Daniele Miorandi	CREATE-NET, Italy
Elisa Marengo	Free University of Bozen-Bolzano, Italy
Emiliano Lorini	IRIT, France
Fabiano Dalpiaz	Utrecht University, The Netherlands
Frank Dignum	Utrecht University, The Netherlands
Guido Governatori	NICTA, Australia
James Cheney	University of Edinburgh, UK
Jordi Sabater Mir	IIIA-CSIC, Spain
Julian Padget	University of Bath, UK
Leon van der Torre	University of Luxembourg, Luxembourg
Liliana Pasquale	The Irish Software Engineering Research Centre, Ireland
M. Birna van Riemsdijk	TU Delft, The Netherlands
Matteo Baldoni	University of Turin, Italy
Nir Oren	University of Aberdeen, UK
Pablo Noriega	Artificial Intelligence Research Institute, Spain
Paolo Torroni	University of Bologna, Italy
Pradeep Murukannaiah	North Carolina State University, USA
Raian Ali	Bournemouth University, UK
Regis Riveret	Imperial College London, UK
Serena Villata	Inria Sophia Antipolis, France
Simon Caton	Karlsruhe Institute of Technology, Germany
Simon Miles	King's College London, UK
The Anh Han	Teeside University, UK
Tina Balke	University of Surrey, UK
Viviana Patti	University of Turin, Italy
Wamberto Vasconcelos	University of Aberdeen, UK

Contents

Automated Negotiation for Traffic Regulation

Matthis Gaciarz[1]([⊠]), Samir Aknine[1], and Neila Bhouri[2]

[1] LIRIS - Université Claude Bernard Lyon 1 - UCBL,
69622 Villeurbanne Cedex, France
matthis.gaciarz@liris.cnrs.fr, samir.aknine@univ-lyon1.fr
[2] IFSTTAR/GRETTIA, Le Descartes 2, 2 rue de la Butte Verte,
93166 Noisy Le Grand Cedex, France
neila.bhouri@ifsttar.fr

Abstract. Urban congestion is a major problem in our society for quality of life and for productivity. The increasing communication abilities of
vehicles and recent advances in artificial intelligence allow new solutions
to be considered for traffic regulation, based on real-time information
and distributed cooperative decision-making models. The paper presents
a mechanism allowing a distributed regulation of the right-of-way of the
vehicles at an intersection. The decision-making relies on an automatic
negotiation between vehicles equipped with communication devices, taking into account the travel context and the constraints of each vehicle.
During this negotiation, the vehicles exchange arguments, in order to
take into account various types of information, on individual and network scales. Our mechanism deals with the continuous aspect of the
traffic flow and performs a real-time regulation.

Keywords: Urban traffic control · Regulation · Negotiation · Cooperative systems · Intersection · Multi-agent system

1 Introduction

Various traffic control methods have been developed in the last decades in order
to optimize the use of existing urban structures. As intersections are conflict
zones causing significant slowdowns, most urban traffic control systems focus on
the intersection regulation, optimizing the right-of-way at traffic lights. Artificial
intelligence enabled to investigate new methods for traffic modeling and regulation, especially with multi-agent technologies, that are able to solve various
problems in a decentralized way [6]. Today's communication technology enables
the design of regulation methods based on real-time communication of accurate
information. Each vehicle on a network has a traffic context, and the information
that constitutes this context can be useful to perform an efficient regulation: the
accumulated delay since the start of the vehicle's journey, its current position,
its short and long-term intentions, etc.

In several countries the rate of vehicles equipped with communication devices,
particularly smartphones, is high, and these devices already change the way

© Springer International Publishing Switzerland 2015
F. Koch et al. (Eds.): CARE-MFSC 2015, CCIS 541, pp. 1–18, 2015.
DOI: 10.1007/978-3-319-24804-2_1

drivers use urban networks by making route recommendation based on real-time information. When numerous vehicles follow these recommandations a traffic reallocation happens. But it is based on the estimation of each vehicle's travel duration and the conflict at intersections is a major source of conflicts and uncertainty. Moreover numerous urban networks are such that bottlenecks cannot be avoided by traffic allocation. Traffic allocation and intersection regulation are complementary aspects and both need to be developed.

Due to the large amount of information, some strategies regulate the traffic on isolated intersection [12]. Some strategies are network-wide control [16] and others focus on the coordination on several intersections creating what is called "green waves" [10]. Green wave reduces stops and gos that cause important time losses. The efficiency of this phenomenon in classical regulation highlights the importance of designing mechanisms enabling coordination at the scale of several intersections. Reference [12] proposes a right-of-way awarding mechanism based on reservation for autonomous vehicles. It relies on a policy called FCFS (First Come First Served), granting the right-of-way to each vehicle asking for it, as soon as possible. This mechanism allows to take into account human drivers by using a classical traffic light policy for human drivers, and giving the right-of-way on red lights to automatic vehicles using the FCFS policy. Although this mechanism accommodates human drivers, its main benefits are due to the FCFS policy and the presence of autonomous vehicles.

In this paper, we propose a different right-of-way awarding mechanism on the intersection scale and tackle two complementary aspects. Firstly, we take into account the traffic context in order to make accurate decisions: the global context (network scale information) and the individual context of each vehicle (history, current information, intentions) are useful information that can be used to produce a fair and efficient regulation policy. Secondly, to have a distributed decision, the vehicles make the decision by themselves in order to deal with the large amount of information. To achieve these goals, we propose a regulation method based on an automatic negotiation mechanism, supported by intelligent agents representing the vehicles' interests. Our mechanism has to bring the vehicles to reach a collective decision in which each vehicle can put forward its individual constraints, suggest solutions and take part in the final decision in real time. Such right-of-way awarding mechanism has to efficiently take into account both autonomous vehicles and human drivers in a vehicle having communication abilities. A fundamental part of our research consists in the conceptualization of multilateral interactions in terms of individual and collective interests. This paper shows a possibility to take some steps towards new foundations of interactions. Based on this, we propose a new negotiation framework for an agent-based traffic regulation and tackle the continuous aspect of the traffic flow. In such negotiations, vehicles build various right-of-way awarding proposals that we call "configurations". These configurations are expounded to the other vehicles of their area, that can raise arguments about the benefits and drawbacks of each configuration. The vehicles decide on the configuration to adopt collectively, with the help of the intersection that contributes to the coordination of the interactions.

The remainder of this article is organized as follows. Section 2 presents the intersection model we opted for, and the problem of right-of-way awarding for an intersection. Section 3 details the method used by agents to build configuration proposals while turning the problem into a CSP (Constraint Satisfaction Problem). Section 4 presents the negotiation mechanism enabling the vehicles to make a collective decision from their individual configuration proposals. It introduces the continuity problem and we detail how the agents tackle it, and presents a complete illustrative scenario. Section 5 gives the experimental results. Finally, Sect. 6 explores future directions and concludes the paper.

2   Problem Description and Intersection Modeling

The problem we are concerned with in this paper is to allocate an admission date to each vehicle arriving at an intersection. This date is defined as a time-slot during which the vehicle has the right-of-way to go into the intersection and cross it. A configuration has to enable an efficient traffic and respect various physical and safety constraints, taking the individual travel context of the vehicles and the global traffic context into account. An agent-based model is used where vehicles and intersections are the agents. The physical representation of the network consists in a cellular automaton model. Cellular automaton models are widely used in literature because they keep the main properties of a network while being relatively simple to use [7]. The intersection is composed of several incoming lanes, called "approaches", and a central zone called "conflict zone". We call "trajectory" the path of a vehicle across the intersection. Each approach and each trajectory is a succession of cells (cf. Fig. 1). A cell out of the conflict zone belongs to exactly one approach. A cell in the conflict zone may belong to one or several trajectories. In this case, this cell is called a "conflict spot".

The moving rules of the vehicles are:

(1) If a vehicle is on the front cell of an approach, this vehicle moves one cell forward and drives into the intersection (the first cell of its trajectory) if and only if it has the right-of-way.
(2) If a vehicle is on an approach, it moves forward if and only if the next cell of the approach is empty, or becomes empty during this time step.
(3) If a vehicle is in the conflict zone, it necessarily moves forward. Our method has to guarantee for each vehicle that it will not meet any other vehicle in the cells of its trajectory.

The decision is distributed: each vehicle agent is able to reason and communicate with the intersection and the other vehicles. To propose a mechanism enabling the vehicles to perform a distributed decision making, the agents may build partial solutions based on their individual constraints, and then merge these partial solutions. Since the admission dates making a configuration are strongly interdependent because of safety constraints, merging partial solutions would be a complex task that would require multiple iterated interactions for the agents with several messages to exchange, and would slow down the decision process.

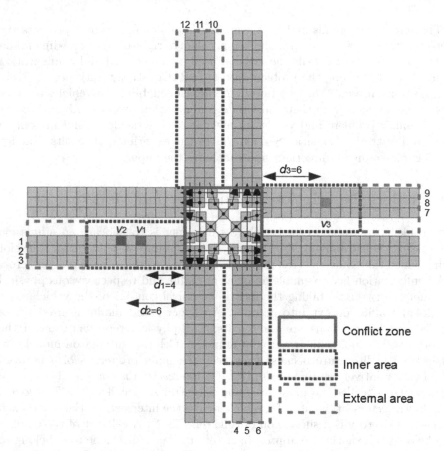

Fig. 1. Intersection with 12 approaches and 12 outcoming lanes, divided into cells. The approaches are numbered from 1 to 12. The conflict zone is crossed by various trajectories, also divided in cells. The cells of the conflict zone are conflict spots. Colored cells are vehicles, e.g. v_1 on the approach 1 is a vehicle coming from the west, about to cross the intersection to the north (Color figure online).

Therefore, in our approach the vehicles build individually full configurations of the intersection and then collectively deliberate on these configurations.

3   Modeling the Right-of-way Allocation Problem to Build Configurations

In order to build configurations, we model the right-of-way allocation problem as a Constraint Satisfaction Problem (CSP) [13]. The CSP fits our problem since it is easy to represent its structural constraints (physical constraints and safety constraints). Let V be the set of all vehicles approaching an intersection, and t_{cur} be the current date in time steps. A configuration is a set $c = \{t_1, ..., t_k\}$ where each t_i is the admission date in the conflict zone accorded to $v_i \in V$. For each $v_i \in V$, app_i is the approach on which is v_i, d_i the distance (in number of cells) between v_i and the conflict zone, $traj_i$ is v_i's trajectory inside the conflict zone. T is the set of all the trajectories inside the conflict zone. $pos(cell_1, traj)$ is the distance, in number of cells, between the cell $cell_1$ and the beginning of the conflict zone on the trajectory $traj$ (the first cell in the conflict zone has the position 0). sp is the speed of the vehicles in cells by time step. In our model, $sp = 1$ cell/time step. We identify 3 types of structural constraints for vehicles, based on the following rules:

R1. Distance rule: A vehicle has to cross the distance separating it from the conflict zone before entering it. We have: $\forall v_i \in V, t_i > t_{cur} + \frac{d_i}{sp}$

R2. Anteriority rule: A vehicle cannot enter the conflict zone before the vehicles preceding it on its lane (this rule could be removed with a more complex model that would take overtaking into account). We have: $\forall v_i, v_j \in V^2, app_i = app_j, d_i < d_j \Rightarrow t_i < t_j$

R3. Conflict rule: Two vehicles cannot be in the same cell at the same time. If the vehicles belong to the same lane or trajectory, the moving rules prevent this case. However, if a cell is a conflict point then we have to model this rule for the vehicles belonging to different trajectories. In a basic version, we have: $\forall v_i, v_j \in V^2, \forall cell_1 \in traj_i, cell_1 \in traj_j \Rightarrow (t_i + \frac{pos(cell_1, traj_i)}{sp}) \neq (t_k + \frac{pos(cell_1, traj_j)}{sp})$. This rule must be reinforced for safety reasons. Indeed, adding a time lapse t_{safe} between the passage of a vehicle on a cell $cell_1$ and the passage of a vehicle in a conflicting trajectory on this cell enhances the drivers' safety (t_{safe} is fixed by an expert). The complete conflict rule is the following:

$\forall v_i, v_j \in V^2, \forall cell_1 \in traj_i, cell_1 \in traj_j \Rightarrow$

$$\left| (t_i + \frac{pos(cell_1, traj_i)}{sp}) - (t_k + \frac{pos(cell_1, traj_j)}{sp}) \right| > t_{safe}$$

A configuration c is valid iff c respects the three rules R1, R2 and R3 and: $\forall v_i \in V, \exists t_i \in c$, where each t_i is v_i's admission date. The scenario represented in Fig. 1 illustrates these three types of structural constraints. Let's consider the three vehicles v_1, v_2, v_3 approaching the intersection at $t_{cur} = 0$. The above rules generate the following 6 constraints:

- **R1** (ct_1) $t_1 > 4$; (ct_2) $t_2 > 6$; (ct_3) $t_3 > 6$
- **R2** (ct_4) $t_2 > t_1$
- **R3** (ct_5) $|(t_1 + 4) - (t_3 + 2)| > 2$; (ct_6) $|(t_2 + 4) - (t_3 + 2)| > 2$

With this CSP model, an agent uses a solver to find compatible admission dates (i.e. respecting the above constraints) for a set $V^{neg} \subseteq V$ of vehicles approaching an intersection. For any configuration c, $\forall v_i \in V^{neg}, \exists d_i \in c$ such as d_i respects the above structural constraints. Several possible configurations may exist for a given situation. A vehicle initially has limited perceptions, however it is able to know in real-time the position of the vehicles around the intersection. As this work conforms the cooperative approach of intelligent transportation systems [2,9], each vehicle has a cooperative behavior with the intersection and communicates its trajectory when it enters the approach of the intersection. With its computation abilities and the available information, a vehicle runs a solver to produce configurations. The use of an objective function enables to guide the CSP solver's search. Moreover, an agent can add additional constraints to its solver as guidelines. If an agent estimates that a particular constraint may produce configurations likely to improve its individual utility or social welfare, this agent considers adding it. However, since this constraint is not a structural constraint resulting from the above rules, it may be violated. The chosen objective function and these potential guideline constraints depend on each vehicle agent's strategy. A configuration built in this manner may satisfy different arguments than the other configurations, and this may be useful in the negotiation to make it chosen.

Example: A bus b and a vehicle v approach an intersection. v and b have conflicting trajectories. Several other vehicles are present on all the approaches of the intersection, so there are numerous structural constraints on the configurations and the search space may be complex to explore. The vehicles consider that buses have priority. v estimates that a good heuristic to find relevant configurations (according to its individual utility and/or social welfare) is to enable a quick admission date to b (below a fixed threshold t_{quick}), and then to search acceptable configurations in this reduced search space. v guides its search by adding to its solver the constraint $t_b \leq t_{quick}$, where t_b is the admission date of b and t_{quick} corresponds to what v considers to be a quick admission date.

4   Right-of-way Negotiation Model

Each vehicle builds configurations allowing it to cross the intersection, however only one configuration will be applied at a given moment. A negotiation process takes place to select it. The mechanism we propose relies on an argumentation-based model [5]. Through the negotiation process, agents aim to reach a collective agreement by making concessions. To perform a negotiation, the vehicle agent relies on its own mental state, made of knowledge, goals and preferences. This mental state evolves during the negotiation. The agents use arguments to make the other agents change their mental states, in order to reach a better compromise. Each agent a_i has the following bases: $\mathcal{K}_i$ is the knowledge base of a_i about its environment. Its beliefs are uncertain, so each belief $k_i^j \in \mathcal{K}_i$ has a certainty level ρ_i^j. $\mathcal{KO}_i$ is the knowledge base of a_i about other vehicles. Each $ko_i^j \in \mathcal{KO}_i$ is a base containing what a_i's believes the knowledge of a_j are. Each of these beliefs has a certainty level δ_i^j. $\mathcal{G}_i$ is the goal base of a_i. These goals have various priority, so each goal $g_i^j \in \mathcal{G}_i$ has a priority level λ_i^j. $\mathcal{GO}_i$ is a_i's base of supposed goals for other vehicles. Each $go_i^j \in \mathcal{GO}_i$ is a base containing what a_i's believes the goals of a_j are. Each of these beliefs has a priority level δ_i^j. Each vehicle has a weight given by the intersections, as detailed in the next section. Two kinds of arguments may be used by the agents, favorable and unfavorable arguments. An argument for (resp. against) a configuration decision d is a quadruple $A = <Supp, Cons, d, w_A>$ where $Supp$ is the support of the argument A, $Cons$ represents its consequences,

w_A is the weight of the argument (fixed by the vehicle v_i that produces this argument and has a weight w_i), such that:

- $d \in \mathcal{D}$, $\mathcal{D}$ being the set of all possible decisions
- $Supp \subseteq \mathcal{K}^*$ and $Cons \subseteq \mathcal{G}^*$
- $Supp \cup \{d\}$ is consistent
- $Supp \cup \{d\} \vdash Cons$ (resp. $\forall g_i \in Cons, Supp \cup \{d\} \vdash \neg g_i$)
- $Supp$ is minimal and $Cons$ is maximal (for set inclusion) among the sets satisfying the above conditions.
- $0 \leq w_A \leq w_i$

Example: A bus b_1 proposes a configuration c_1 allowing it to cross the intersection as quick as possible to catch up its lateness. A vehicle v_1 precedes this bus on the same lane. Giving a quick admission date to b_1 (below a fixed threshold t^b_{quick}) implies to give a quick admission date to v_1 (below a fixed threshold t^v_{quick}), and one of the goals of v_1 is to cross the intersection as quick as possible. Thus:

$\mathcal{K}_{v_1} = \{crossesQuickly(b_1) \rightarrow crossesQuickly(v_1)\}$
$\mathcal{G}_{v_1} = \{crossesQuickly(v_1)\}$
v_1 may take advantage of this configuration, so it produces the following argument:

$< \{crossesQuickly(b_1), crossesQuickly(b_1) \rightarrow crossesQuickly(v_1)\},$
$\{crossesQuickly(v_1)\}, c_1 >.$

For safety reasons, the intersection has a current configuration at any time. The goal of an agent through the negotiation is to change this current configuration c_{cur} by another c_{best} that improves its individual utility. In a negotiation the agents rely on a communication language to interact. The set of possible negotiation speech acts is the following: $Acts = \{Offer, Argue, Accept, Refuse\}$.

Offer(c_{new}, c_{cur}): with this move, an agent proposes a configuration c_{new} to replace c_{cur}. An agent can only make each offer move once.

Argue$(c, arg(c))$: with this move, an agent gives an argument in favor of c or against c.

Accept(c_{new}, c_{cur}), **Refuse**(c_{new}, c_{cur}): with these moves, an agent accepts (resp. refuses) a configuration c_{new} to replace c_{cur}.

c_{new} is accepted iff $\frac{\sum_{v_i \in V(c_{new})} w_i}{\sum_{v_i \in V^{neg}} w_i} \geq th_{accept}$, where:

th_{accept} is an acceptance threshold ($th_{accept} > 0.5$).
$V(c_{new}) \subseteq V^{neg}$ is the set of vehicles accepting the configuration $c_{new} \in \mathcal{D}$ to replace c_{cur}. w_i is a weight given by the intersections to the vehicle v_i. When a configuration is adopted by the agents, this configuration becomes the current configuration of the intersection (Fig. 2).

4.1   Role of the Intersection Agent

In order to perform a right-of-way allocation that maximizes the social welfare and encourages cooperative behaviors, the intersection agent takes part in the

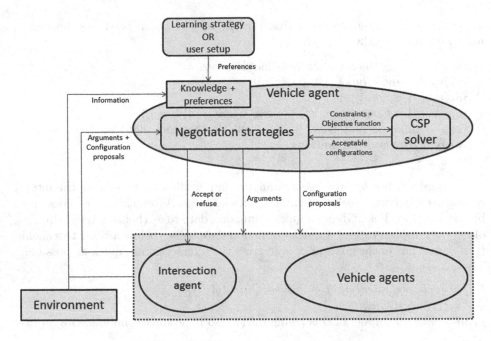

Fig. 2. Vehicle agent and negotiation

negotiation process. Each vehicle first defends its own interests, and also defends other interests that may guide the negotiation towards a favorable outcome for it. A vehicle can represent the interests of other vehicles outside V^{neg} (for example the vehicles that follow it) or network scale interests (for example clearing some lanes) if it can get advantage of it. However, it may happen that these arguments do not directly concern the vehicles of V^{neg}, that may ignore these arguments despite their positive contribution to global social welfare. To avoid this effect, the intersection agent is able to represent these external interests. Like the vehicle agents, the intersection agent has its own mental states and is able to produce arguments. However, it cannot accept or refuse proposals.

The weight the intersection agent gives to each of its arguments depends on the importance of the external interests represented by these arguments. A weight w_i of a vehicle v_i is given by the intersection agents to encourage the vehicles to have cooperative behaviors. According to v_i's cooperation level in its negotiation behavior, the intersection increases or decreases w_i for the remainder of v_i's journey. A vehicle refusing a proposal having numerous strong arguments for it (or accepting a proposal having numerous strong arguments against it) gets an important weight penalty. On the contrary, a vehicle accepting a proposal having numerous strong arguments for it (or refusing a proposal having numerous strong arguments against it) gets a weight reward. For a vehicle, these rewards and penalties are significant in the middle and long term since it affects durably its capacity to influence the choice of the configurations on the next intersections. To perform this, the intersection uses arguments to assign a reward (or penalty)

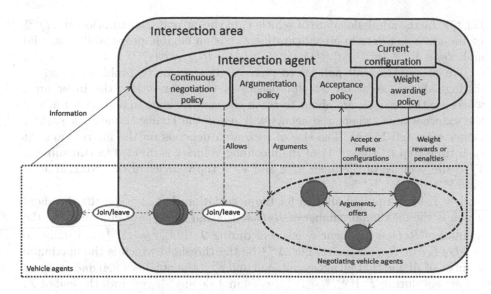

Fig. 3. Role of the intersection agent

value to each proposal, so that the vehicles may evaluate the benefits and risks
from each decision about configurations before making it.

The intersection uses reward or penalty according to the weight of the vehicles. A vehicle that already has a high weight gets a little advantage while getting
a weight reward, but getting a weight penalty would be an important drawback.
On the contrary, a vehicle having a low weight would get a little drawback
from a weight penalty and an important advantage from a weight reward. Let
$V^{min} \in V^{neg}$ be the set of the vehicles that emitted arguments contradictory
to the intersection agent's preference. To have more influence on the vehicles,
the intersection agent uses penalties when the average weight of the vehicles of
V^{min} is greater than the average weight of the vehicles of V^{neg}, and uses rewards
otherwise (Fig. 3).

4.2 Continuous Negotiation Mechanism

Since the flow of vehicles is continuous, the mechanism has to manage this
dynamic aspect by defining the agents that take part in each negotiation step,
the vehicles for which this configuration provides an admission date, and the conditions under which this configuration could be revised once chosen. In order to
manage technical failures, the intersection has a current configuration c_{cur} at any
time. According to the chosen continuity policy, the negotiation mechanism may
allow the vehicles to collectively change this configuration. However, the mechanism has to consider safety measures before allowing this change. Changing the
configuration at the last moment is risky because of the slowness of the reaction
of the drivers. To avoid this, we define a safety time threshold th_{safe}. The admission date of a vehicle cannot be revised (removed or granted) in a too short term.

Let t_i^{cur} be the admission date of vehicle v_i in the current configuration and t_i^{next} be its admission date in a configuration c. c is an eligible proposal iff c is valid and: $\forall v_i \in V^{neg}, (t_i^{cur} = t_i^{next}) \vee ((t_i^{cur} \geq t_{cur} + th_{safe}) \wedge (t_i^{next} \geq t_{cur} + th_{safe}))$.

We propose several policies to manage the continuity problem. First, we distinguish two areas on the approaches of the intersection: the inner area, where all the vehicles are about to reach the conflict zone in a short term, and the external area, where the agents will reach the conflict zone in a slightly longer term (cf. Fig. 1). The size of each area depends on the intersection. At each time step t_i, the set V_i of the incoming vehicles is divided in two subsets: V_i^{inn} the vehicles of the inner area and V_i^{ext} the vehicles of the external area. $V_i = V_i^{inn} \cup V_i^{ext}, V_i^{inn} \cap V_i^{ext} = \emptyset$

Let $\mathcal{T}$ be the period allowed for the negotiation. Let Δ^{ref} be the threshold which is the maximum number of $Refuse$ that an agent can send and δ_i^{ref} the number of $Refuse$ an agent v_i has sent during $\mathcal{T}$. If $\delta_i^{ref} = \Delta^{ref}$, v_i cannot do any $Offer$ or $Refuse$ move. Let Δ^{arg} be the threshold which is the maximum number of $Argue$ that an agent can send and δ_i^{arg} the number of $Argue$ an agent v_i has sent during $\mathcal{T}$. If $\delta_i^{arg} = \Delta^{ref}$, v_i cannot do any $Argue$ until the end of $\mathcal{T}$. An agent can only make each offer once during a negotiation. Once an agent has made the move $Offer(c_x, c_y)$ during $\mathcal{T}$, it cannot make it again during the negotiation. We get the following set of rules.

- **NR1:** $\forall v_i \in V^{neg}$, the move $Offer(c_x, c_y)$ can be made at any time by v_i if this move has not been made yet by v_i during $\mathcal{T}$ and if $\delta_i^{ref} < \Delta^{ref}$.
- **NR2:** $\forall v_i \in V^{neg}$, the move $Accept(c_x, c_y)$ can be made at any time by v_i. Furthermore, the move $Offer(c_x, c_y)$ was made at time $t_0 \in \mathcal{T}, t_0 < t$.
- **NR3:** $\forall v_i \in V^{neg}$, the move $Refuse(c_x, c_y)$ can be made at any time $t \in \mathcal{T}$ by v_i if $\delta_i^{ref} < \Delta^{ref}$. Furthermore, the move $Offer(c_x, c_y)$ was made at time $t_0 \in \mathcal{T}, t_0 < t$.
- **NR4:** $\forall v_i \in V^{neg}$, the move $Argue(c_x, arg(c_x))$ can be made at any time $t \in \mathcal{T}$ by v_i if $\delta_i^{arg} < \Delta^{arg}$. Furthermore, the move $Offer(c_x, c_y)$ was made at time $t_0 \in \mathcal{T}, t_0 < t$, for any $c_y \in \mathcal{D}$.

Iterated Policy (IP). With this policy, the vehicle agents join the negotiation by waves, and perform iterated decisions that cannot be revised. At a given instant t_{i-1}, V^{inn} is empty. At the next time step t_i, since the vehicles have moved, V^{inn} and V^{ext} change. The set of negotiating vehicles V_i^{neg} becomes equal to V_i^{inn}. Then the vehicles of V_i^{neg} perform a collective decision about the configuration for all the vehicles of V_i^{neg}. A negotiation process starts, with a limited duration d_{neg} in addition to the above set of rules. $\mathcal{T} = [t_0^{neg}, t_0^{neg} + d_{neg}]$, where t_0^{neg} is the starting date of the negotiation. With this limited duration, the agents have interest to quickly make reasonable proposals for every vehicle. At the end of this negotiation step, a configuration c_i is chosen, awarding an admission date to each vehicle of V_i^{neg}.

At t_{i+1}, a new iteration begins, and $V_{i+1}^{neg} = V_{i+1}^{inn} \setminus V_i^{neg}$. The vehicles of V_{i+1}^{neg} start a new negotiation, but the vehicles that already have taken part in a previous negotiation step do not take part in this one. The agents of V_{i+1}^{neg} are not allowed to revise c_i, the agents only negotiate the admission dates of the vehicles of V_{i+1}^{neg} since the other vehicles of V_i^{inn} already have an admission date defined in c_i or in previous configurations. A new configuration c_{i+1} is chosen, similar to c_i except it adds admission dates for the vehicles of V_{i+1}^{neg}. $c_i \backslash c_i^{out} \subseteq c_{i+1}$ where c_i^{out} is the set of the vehicles admitted in the conflict zone: $\forall t_j \in c_i, t_j < t_i \Leftrightarrow t_j \in c_i^{out}$.

The policy continues to iterate and to produce new admission dates for the next vehicles in the inner area without revising those of the vehicles that already were in it.

An extended policy EIP (Extended Iterated Policy) has been defined from IP. This policy is similar to IP, except that whenever an iteration ends, the new iteration does not necessarily start straightaway. If $V_i^{neg} = V_i^{inn} \backslash V_{i-1}^{neg}$ only contains a few low-weighted vehicles, it is better to wait before starting a new negotiation step. In this case, the intersection gives a temporary admission date to the vehicles of V_i^{neg} using the FCFS (First Come First Served) policy on the current configuration. Extending a configuration with FCFS consists in granting to each new vehicle the first available admission date, without changing the admission dates of the previous vehicles. These vehicles take part in the next negotiation iteration and can revise their temporary admission date. In this case, $V_{i+1}^{neg} = V_{i+1}^{inn} \backslash V_{i-1}^{neg}$.

Continuous Policy (CP). When this policy is applied the vehicles dynamically join the current negotiation while entering the inner area, $V^{neg} = V^{inn}$ at any time. When a vehicle v_{new} joins V^{inn}, all the useful information about the current state of the negotiation (configurations and arguments) are communicated to v_{new} so that it can join the negotiation. The current configuration of the intersection can be totally revised by a collective decision, except for the vehicles that are concerned by the security threshold.

Whenever new vehicles join V^{inn}, the current configuration of the intersection and the configurations under negotiation do not provide admission dates for these vehicles, since the configurations were emitted before these vehicles joined V^{inn}. However, the intersection provides an ordering on these vehicles. With this ordering, it is possible for any vehicle in the negotiation to extend any of the vehicles' configuration proposal. Extending a configuration consists in adding an admission date for each new vehicle with the FCFS strategy, using the ordering on these vehicles. The agents consider that any proposal in the negotiation that do not provide an admission date to each vehicle of V^{inn} will be extended with FCFS. It guarantees that the intersection always has an admission date for each vehicle of V^{inn}. Thus, even if the negotiation always fails, the FCFS policy is applied.

A possible perspective is to extend CP with a new policy CPA (Continuous Policy with Anticipation). In CP, when a vehicle builds a configuration, this configuration only incorporates vehicles of V^{inn}. In CPA, each vehicle $v_1 \in V^{neg}$ can take into account any other vehicle from $v_2 \in V^{ext}$ while building configurations, in order to take advantage of it. Then, whenever v_2 joins V^{inn}, some proposals (including the current configuration of the intersection) may already include an admission date for it. According to the result of the previous negotiations these configurations may be better than the one produced by the FCFS strategy.

4.3   Illustrative Scenario

We continue the scenario described in Sect. 3 (Fig. 1). Each vehicle has built the structural constraints to model the problem and has run a solver to build configurations. Three Pareto-optimal configurations are possible: $c_1 = \{5, 7, 12\}$, $c_2 = \{5, 11, 10\}$, $c_3 = \{8, 9, 7\}$. For instance, the admission date of v_1 in configuration c_1 is $t_1^{c_1} = 5$. On a very simple scenario like this one, we can easily assume that each vehicle's solver produces these 3 configurations during its first search, and even other suboptimal solutions. But when the number of vehicles approaching the intersection is high, the search space is very large and all vehicles will not necessarily find all the Pareto-optimal solutions. To illustrate this phenomenon, let's assume that all the vehicles do not find the 3 Pareto-optimal solutions during their first search. Let's also assume that results of the first search give the following configurations: (c_2, c_3) for v_1, c_3 for v_2, and (c_1, c_2) for v_3.

The initial context is the following: the intersection has applied a FCFS policy to compute a default configuration, so the current configuration c_{cur} is $c_2 = \{5, 11, 10\}$. v_3 has a cooperative behavior since the beginning of its travel so it now has a higher weight than the two other vehicles: $w_1 = 10, w_2 = 10, w_3 = 25$. We assume that an important group of vehicles gr_1 is incoming on v_3's lane, and the sum of the weights of these vehicles is $w_{gr} = 40$. The acceptance threshold $th_{accept} = 0.5$. The mental states of the agents are given in Table 1. The goals in this table can be either produced by a learning system or set up by the user. The agents have three types of goals. (1) With goal $improve(v_i)$, the agent aims to improve v_i's admission date, in order to cross the next intersection as soon as possible. (2) With $group(v_i)$ the agent aims to make v_i form a physical group,

Table 1. Initial mental states

v_i	$\mathcal{K}_i$	$\mathcal{G}_i$
v_1	$\mathcal{K}_1 = \{k_1^1 = (t_1^{new} < t_1^{cur} \rightarrow improve(v_1))(\rho_1^1 = 1)\}$	$\mathcal{G}_1 = \{g_1^1 = improve(v_1)(\lambda_1^1 = 0.4),$ $g_1^3 = weight(v_1)(\lambda_1^3 = 0.6)\}$
v_2	$\mathcal{K}_2 = \{k_2^1 = (t_2^{new} < t_2^{cur} \rightarrow improve(v_2))(\rho_2^1 = 1), k_2^2 = (t_2^{new} - t_1^{new} \leq 3 \rightarrow group(v_2)(\rho_2^2 = 1)\}$	$\mathcal{G}_2 = \{g_2^1 = improve(v_2)(\lambda_2^1 = 0.3),$ $g_2^2 = group(v_2)(\lambda_2^2 = 0.4),$ $g_2^3 = weight(v_2)(\lambda_2^3 = 0.4)\}$
v_3	$\mathcal{K}_3 = \{k_3^1 = (t_3^{new} < t_3^{cur} \rightarrow improve(v_3))(\rho_3^1 = 1)\}$	$\mathcal{G}_3 = \{g_3^1 = improve(v_3)(\lambda_3^1 = 0.8),$ $g_3^3 = weight(v_3)(\lambda_3^3 = 0.2)\}$
v_4	$\mathcal{K}_4 = \{k_4^1 = (t_4^{new} < t_4^{cur} \rightarrow improve(v_4))(\rho_4^1 = 1), k_4^2 = (t_4^{new} - t_3^{new} \leq 3 \rightarrow group(v_4)(\rho_4^2 = 1)\}$	$\mathcal{G}_4 = \{g_4^1 = improve(v_4)(\lambda_4^1 = 0.3),$ $g_4^2 = group(v_4)(\lambda_4^2 = 0.5),$ $g_4^3 = weight(v_4)(\lambda_4^3 = 0.2)\}$

Table 2. Negotiation process

Step	t_{cur}	v_1	v_2	v_3	v_4	it	c_{cur}	Move(s)
	0	$c_2 \succ c_3$	c_3	$c_2 \succ c_1$		c_2	c_2	
0		-	$c_3 \succ c_2$	-		-	c_2	$v_2 : m_1$
1		-	-	$c_3 \succ c_2 \succ c_1$		$c_2 \sim c_3$	c_2	$v_1 : m_2;\ v_2 : m_3;$ $v_3 : m_4$
2		-	-	-		$c_3 \succ c_2$	c_2	$it : m_5$
3		$c_3 \succ c_2$	-	-		-	c_2	$v_1, v_2, v_3 : m_6$
4	1	$c_1 \succ c_3 \succ c_2$	-	-		-	c_3	$v_1 : m_7$
5		-	$c_1 \succ c_3 \succ c_2$	-		$c_3 \succ c_2 \sim c_1$	c_3	$v_1 : m_8;$ $v_2 : m_9, m_{10};$ $v_3 : m_{11};\ it : m_{12}$
6		-	-	-		$c_1 \succ c_3 \succ c_2$	c_3	$it : m_{13}$
7		-	-	-		-	c_3	$v_1, v_2 : m_{14};$ $v_3 : m_{15}$
8	2	-	-	-	$c_1 \succ c_3$	-	c_3	$v_4 : m_{16}$
9		-	-	$c_1 \succ c_3 \succ c_2$	-	-	c_3	$v_3, v_4 : m_{14}$
10	3	-	-	-	-	-	c_1	...

Table 3. Argumentation moves used in the negotiation

Move name	Move description	Positive argument	Argument
m_1	$Offer(c_3, c_2)$	/	/
m_2	$Argue(c_3, Arg_1)$	no	$Arg_1 =< \{t_1^{c_3} \geq t_1^{cur}, t_1^{c_3} \geq t_1^{cur} \rightarrow \neg improve(v_1)\}, \{improve(v_1)\}, c_3, w_1 >$
m_3	$Argue(c_3, Arg_2)$	yes	$Arg_2 =< \{t_2^{c_3} < t_2^{cur}, t_2^{c_3} < t_2^{cur} \rightarrow improve(v_2)\}, \{improve(v_2)\}, c_3, w_2 >$
m_4	$Argue(c_3, Arg_3)$	yes	$Arg_3 =< \{t_3^{c_3} < t_3^{cur}, t_3^{c_3} < t_3^{cur} \rightarrow improve(v_3)\}, \{improve(v_3)\}, c_3, w_3 >$
m_5	$Argue(c_3, Reward_1)$	yes	$Reward_1 =< \{weight(any)\}, \{weight(any)\}, c_3, w_2 + w_3 - w_1 >$
m_6	$Accept(c_3, c_2)$	/	/
m_7	$Offer(c_1, c_3)$	/	/
m_8	$Argue(c_1, Arg_4)$	yes	$Arg_4 =< \{t_1^{c_1} < t_1^{cur}, t_1^{c_1} < t_1^{cur} \rightarrow improve(v_1)\}, \{improve(v_1)\}, c_1, w_1 >$
m_9	$Argue(c_1, Arg_5)$	yes	$Arg_5 =< \{t_2^{c_1} < t_2^{cur}, t_2^{c_1} < t_2^{cur} \rightarrow improve(v_2)\}, \{improve(v_2)\}, c_1, w_2 >$
m_{10}	$Argue(c_1, Arg_6)$	no	$Arg_6 =< \{t_3^{c_1} \geq t_3^{cur}, t_3^{c_1} \geq t_3^{cur} \rightarrow \neg improve(v_3)\}, \{improve(v_3)\}, c_1, w_3 >$
m_{11}	$Argue(c_1, Arg_7)$	yes	$Arg_7 =< \{t_2^{c_1} - t_1^{c_1} \leq 3, t_2^{c_1} - t_1^{c_1} \leq 3 \rightarrow group(v_2)\}, \{group(v_2)\}, c_1, w_2 >$
m_{12}	$Argue(c_1, Arg_8)$	yes	$Arg_8 =< \{t_3 > 10, t_3 > 10 \rightarrow group(gr_1)\}, \{group(gr_1)\}, c_1, w_{gr} >$
m_{13}	$Argue(c_1, Threat_1)$	no	$Threat_1 =< \{\neg weight(any)\}, \{weight(any)\}, c_1, w_1 + w_2 + w_{gr} - w_3 >$
m_{14}	$Accept(c_1, c_2)$	/	/
m_{15}	$Refuse(c_1, c_2)$	/	/
m_{16}	$Argue(c_1, Arg_{11})$	yes	$Arg_{11} =< \{t_4^{c_1} - t_3^{c_1} \leq 3, t_4^{c_1} - t_3^{c_1} \leq 3 \rightarrow group(v_4)\}, \{group(v_4)\}, c_1, w_4 >$

called "platoon", with other vehicles on the same lane. Vehicles forming a platoon often have common interests and may naturally have a common negotiation behavior at the intersections. Such behavior gives them a high weight in the negotiations, and brings an important advantage on the long range. This goal represents the desire of the vehicles to form platoons in order to get advantage of this potential phenomenon. (3) With $weight(v_i)$ the agent aims to keep v_i's weight high enough to be influential in negotiations at the next intersections. For sake of simplicity, the evaluation of these goals can only get a boolean value: achieved or not achieved. The negotiation is described in Table 2. This table gives the preferences of agents v_1, v_2, v_3, and the intersection agent it. $c_x \succ c_y$ means that c_x is preferred to c_y. $c_x \sim c_y$ means that the agent is indifferent between c_x and c_y. '-' means that the preferences of the agent have not changed since the previous step. During each negotiation step, the agents produce negotiation moves described in Table 3.

v_2 can improve its admission date with c_3 and offers it to the negotiation (step 0). c_3 improves v_2 and v_3's admission dates and deteriorates v_1's date. v_2 and v_3 build positive arguments on c_3, and v_1 builds a negative argument (step 1). The positive arguments are stronger than the negative one, so the intersection rewards the vehicles that would vote for c_3, with a weight equal to the relative strength of the arguments (step 2). The reward is high enough to change v_1's preferences, and v_1 accepts c_3. v_2 and v_3 are favorable to c_3 and accept it. All the vehicles accepted c_3, so it replaces c_2 as the current configuration of the intersection (step 3).

The negotiation continues. A time step elapsed since the beginning of the negotiation, during which v_1's solver has found c_1. Since c_1 is now v_1's preferred solution, v_1 offers c_1 (step 4). The vehicle agents give their arguments for c_1 (v_1 and v_2) or against c_1 (v_3). The intersection agent estimates that the vehicles of gr_1 can get advantage of c_1, so it gives a new argument for c_1 based on this information, with a weight equal to w_{gr} (step 5). The negative arguments are stronger than the positive one, so the intersection threats the vehicles that would vote for c_1, with a weight equal to the relative strength of the arguments (step 6). The penalty is not high enough to change v_3's preferences, and v_3 refuses c_1. v_1 and v_2 were favorable to c_3, but their cummulated weight is not high enough to change the configuration, and c_3 remains the current configuration of the intersection. v_3 is threatened and if it does not change its refusal into an acceptance before crossing the intersection, its weight will be reduced (step 7).

The negotiation continues. At the next time step, a new vehicle $v_4 \in gr$ enters the inner area. In this scenario, the continuous policy is applied. Since v_4 joins v_{inn}, it immediately joins the negotiation. Its individual weight is $w_4 = 10$. v_4 gets all the negotiation information, and its admission date is added to each configuration with FCFS. We now have: $c_1 = \{5, 7, 12, 13\}$, $c_2 = \{5, 11, 10, 16\}$, $c_3 = \{8, 9, 7, 14\}$. c_1 is v_4's preferred solution so v_4 gives a new argument for it (step 8). Since the total weight of the vehicles that prefer c_1 over c_3 is greater than the weight of the vehicles that prefer c_3 over c_1, v_3 risks a weight penalty without any reward if it does not change its refusal into an acceptance, so it accepts c_1. Moreover, c_1 is v_4's preferred configuration. v_3 and v_4 accept c_1 (step 9). c_1

replaces c_3 as the current configuration of the intersection (step 10). The negotiation continues continuously, step by step.

5 Experimentation and Discussion

This work has been implemented in Java with the Choco library for CSP [8], on an intersection with 12 approaches (cf. Fig. 1). The length of the inner area is 6 cells on each approach. Agents are implemented as threads: each agent has its own solver and its own negotiation strategy. The agents communicate with other agents with direct messages. On a personal computer (RAM 2 Gb, 1.9 GHz mono-core processor), 2 s are enough to run the solver and compute

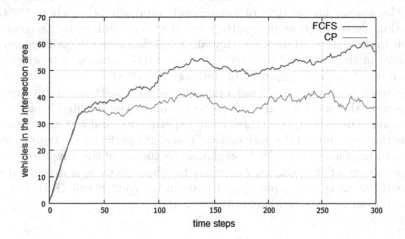

Fig. 4. Number of vehicles in the area

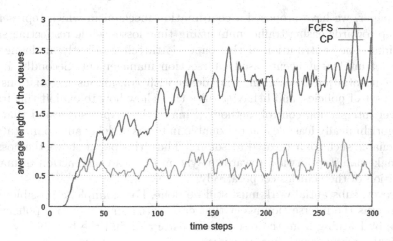

Fig. 5. Average length of the queues

several good configurations for about 30 vehicles, and the negotiation time is low enough to enable to run the mechanism in real time. In this section, we present the results of the comparison between FCFS and the CP policy. We simulated a continuous incoming flow of vehicles (1.2 vehicle/step in average). Vehicles appear on a randomly chosen lane with stochastically generated goals and knowledge. $n_{arg} = 10$ kinds of goals are provided. Each vehicle has a priority level for each of these, and their sum is normalized to 1. Then each vehicle is able to generate up to a maximum of 3 stochastic arguments for each configuration. Each argument supports one of the goals and has a random weight (that may be positive or negative). We chosed to apply $t_{safe} = 2$. These simulations were performed on a more powerful computer with RAM 32 Gb, 64-core processor. Results are shown on Figs. 4 and 5 for 20 simulations. These figures respectively represent the number of vehicles in the intersection area and the average number of vehicles waiting for the right of way on each approach, relatively to the time. Simulations have a 300 steps length, each step representing one second. For example in simulations of the CP policy, after 100 time steps the average number of vehicles in the area were 37.9 (cf. Fig. 4) and 0.64 vehicles were waiting for the right of way on each approach of the intersection (cf. Fig. 5).

The main improvements of our negotiation-based mechanism are expected to appear on the network scale, and so far we only experimented it on a single intersection. The main goal of these early experiments, and our main result, is to show the feasability of this mechanism. The slight performance improvements shown on Figs. 4 and 5 may also be explained by the use of the solver to optimise the right-of-way of the vehicles. Moreover, this improvement is accentued with the use of the safety time lapse t_{safe} defined in the conflict rule (R3) that gives more importance to the ordering of the vehicles.

6   Conclusion

In this paper, we have proposed a coordination mechanism which represents a large step towards easing traffic, minimizing time losses while respecting safety constraints. The contribution of this paper is threefold. Firstly, it defined the problem of intelligent agent-based intersection management. Secondly, it presented a negotiation mechanism that deals with continuous negotiations and applies a set of policies, and behavior rules that show how to exploit this framework over intersection control methods. Finally this paper suggested that it is both algorithmically feasible and reasonable in terms of delay and computational cost to enable such sophisticated reasoning. Thus, this paper shows the possibility to make one step forward towards a system that can take action to manage the decision of the vehicles cooperatively.

However, substantial work must still be done. For example, a possible direction concerns the intersection agent that can switch among several policies, for instance by learning from the reservation history to find the best policy suited to particular traffic conditions. In current work we are adapting the behavior of the intersection to handle vehicle priorities.

Finally, an interesting direction would be to make a link with traffic allocation problem. As explained in the introduction, single intersection regulation and traffic allocation are complementary problems, and it would be relevant to consider how some aspects of each problem can be taken into account in the other. For example an anticipated negotiation of the right-of-way would allow to make a precise estimation of the waiting time of a vehicle at an intersection, that may lead it to revise its itinerary. Moreover a negotiation mechanism similar to the one presented in this paper may allow important groups of vehicles to negotiate both their long-term itinerary and the right-of-way for the intersection on this itinerary.

Acknowledgments. Funding for this project was provided by a grant from la Région Rhône-Alpes. The authors would like to acknowledge the students Guillaume Collombet, Paul Talvat, Anita Barry, Bruno Dumas, Loubna Elmanany, Jérémy Ferrer, Damien Mornieux and Antoine Richard for their support in the implementation of the mechanism.

References

1. Abbas-Turki, A. et al.: Cooperative intersections for emerging mobility systems. In: 15th Euro Working Group on Transportation (2012)
2. Abbas-Turki, A., Perronnet, F., Buisson, J., El-Moudni, A., Ahmane, M., Zéo, R., et al.: Cooperative intersections for emerging mobility systems. In: 15th Euro Working Group on Transportation (2012)
3. Adler, J.L., et al.: A multi-agent approach to cooperative traffic management and route guidance. Transp. Res. **39**, 297–318 (2005)
4. Adler, J.L., Satapathy, G., Manikonda, V., Bowles, B., Blue, V.J.: A multi-agent approach to cooperative traffic management and route guidance. Transp. Res. Part B **39**(4), 297–318 (2005)
5. Amgoud, L., Belabbes, S., Prade, H.: Towards a formal framework for the search of a consensus between autonomous agents. In: Parsons, S., Maudet, N., Moraitis, P., Rahwan, I. (eds.) ArgMAS 2005. LNCS (LNAI), vol. 4049, pp. 264–278. Springer, Heidelberg (2006)
6. Bazzan, A.L.C., Klügl, F.: A review on agent-based technology for traffic and transportation. Knowl. Eng. Rev. **29**(03), 375–403 (2013)
7. Brockfeld, E., Barlovic, R., Schadschneider, A., Schreckenberg, M.: Optimizing traffic lights in a cellular automaton model for city traffic. Phys. Rev. E **64**, 056132 (2001)
8. T. choco team: choco: an open source java constraint programming library (2010)
9. Choy, M.C., Srinivasan, D., Cheu, R.L.: Cooperative, hybrid agent architecture for real-time traffic signal control. IEEE SMC **33**(5), 597–607 (2003)
10. de Oliveira, D., Bazzan, A., Lesser, V.: Proceedings of the Fourth International Joint Conference on Autonomous Agents and Multiagent Systems, pp. 463–470. ACM (2005)
11. Dresner, K., Stone, P.: Sharing the road: autonomous vehicles meet human drivers. In: IJCAI (2007)
12. Dresner, K., Stone, P.: A multi-agent approach to autonomous intersection management. J. Artif. Intell. Res. **39**, 591–656 (2008)

13. Kumar, V.: Algorithms for constraint-satisfaction problems: a survey. AI Mag. **13**(1), 32–44 (1992)
14. Maerivoet, S., De Moor, B.: Cellular automata models of road traffic. Phys. Rep. **419**, 1–64 (2005)
15. Monteil, J., Billot, R., El Faouzi, N.-E.: Towards cooperative traffic management: methodological issues and perspectives. In: Proceedings of Australasian Transport Research Forum 2011, Adelaide, Australia (2011)
16. Roozemond, D.: Using intelligent agents for proactive, realtime urban intersection control. Eur. J. Oper. Res. **131**, 293–301 (2001)

Towards a Middleware for Context-Aware Health Monitoring

Eduardo A. Oliveira[1]([⊠]), Fernando Koch[2], Michael Kirley[1],
and Carlos Victor G. dos Passos Barros[3]

[1] Department of Computing and Information Systems,
The University of Melbourne, Parkville, VIC 3010, Australia
{eduardo.oliveira,mkirley}@unimelb.edu
[2] SAMSUNG Research Institute, São Paulo, Brazil
fernando.koch@samsung.com
[3] Recife Center for Advanced Studies and Systems, Recife, Brazil
carlos.gomes@cesar.org.br

Abstract. The surge of commodity devices, sensors and apps allows for
the continuous monitoring of patient's health status with relatively low-
cost technology. Nonetheless, current solutions focus on presenting data
and target at individual health metrics and not intelligent recommendations. In order to advance the state-of-the-art, there is a demand for
models that correlate mobile sensor data, health parameters, and situational and/or social environment. We seek to improve current models by
combining environmental monitoring, personal data collecting, and predictive analytics. For that, we introduce a middleware called Device Nimbus that provides the structures to integrate data from sensors in existing
mobile computing technology. Moreover, it includes the algorithms for
context inference and recommendation support. This development leads
to innovative solutions in continuous health monitoring, based on recommendations contextualised in the situation and social environment.
In this paper we propose a model, position it against state-of-the-art,
and outline a proof-of-concept implementation.

Keywords: Intelligent agent · Context aware · Health · Middleware

1 Introduction

Wearable Health-Monitoring Systems is receiving large attention by both industry and academic research [7,11,18]. Current solutions focus on collecting mobile
sensor data and presenting data and target at individual health metrics. They
fail in proposing intelligent recommendations and correlating with situational
and/or social environment. For instance, an application that measures heart-
beat rate issues an alarm if a threshold is reached and the user is running.
However, it does not take into consideration that the user is running at a park
with a colleague. In this case, it could issue the warning to his running mate
to slow the pace, thus enhancing the recommendation efficiency. Hence, there is

© Springer International Publishing Switzerland 2015
F. Koch et al. (Eds.): CARE-MFSC 2015, CCIS 541, pp. 19–30, 2015.
DOI: 10.1007/978-3-319-24804-2_2

a demand for a solution able to integrate data from multiple sources and support advanced decision making in automated health recommendation, based on contextual and social aspects.

Context awareness provides the tools for personalised health monitoring. It provides techniques to implement noise filtering, information selection, and service adaptiveness [8]. For that, accurate inference of context parameters is paramount to support alarming and intelligent recommendation in health monitoring. However, we identified a lack of techniques to infer context parameters based on social health aspects. We seek to improve current models of data aggregation from multiple sources, social data, situation data, and predictive analytics. This development will support innovative solutions in health monitoring that relate situational and/or social environment to provide recommendations and decision support.

We introduce a middleware called "Device Nimbus" that provides the structures to integrate data from diverse sensors in commodity mobile computing technology and execute the models of context and predictive analysis. The solution is being designed to fulfilling the requirements of Internet of Things, namely: heterogeneity, e.g. different sensors, protocols and applications; dynamicity, e.g. arrival and departure of devices and sensors; analysis, e.g. contents personalization, recommendations and prediction, and; evolution, e.g. support for new protocols, devices and sensors. The proposal encompasses three main components: Data Collectors, Data Integration and Intelligent Modules. The resulting solution addresses the requirements of the target scenario by providing context-awareness, adaptivity, flexibility and extensibility to the proposed middleware.

This paper is organised as follows. Section 2 details the proposal for "Device Nimbus", presenting requirements and expected results. Section 3 provides an overview of the state-of-the-art and comparative analysis. The paper concludes with Sect. 4 by providing our perspective on technology development and future work.

2   Proposal

Device Nimbus provides a stepping-stone towards solving many of the problems currently found in health domain such as: tracking users based in lots of different mobile devices/sensors/protocols, providing personalized feedbacks and getting connected with clinics and hospitals and, was designed to deal with big amounts of data. Many people agree that middleware plays a vital role in hiding the complexity of distributed applications. Middleware typically operate in an environment that may include heterogeneous computer architectures, operating systems, network protocols, devices and databases [15]. Device Nimbus middleware will progress the state of the art supporting the design of health systems and applications composed of a large number of independent, autonomous, heterogeneous and interacting sub-systems, sensors and mobile devices. Developers will be able use this middleware when developing new apps, which can collect and analyze personal metrics/ data in a variety of pre-determined ways.

Network communication, scalability, reliability, coordination and heterogeneity have been common requirements of traditional middleware such as Remote Procedure Calls (RPC), Message Oriented Middleware (MOM), Distributed Computing Environment (DCE), Transaction Processing Monitors (TPMON), and Object Oriented Middleware (ORB) [15]. Therefore, some requirements different from traditional middleware have to be considered for the middleware supporting applications and services in the ubiquitous environment. We propose the following new requirements for the platform:

- Context-Awareness: context should include device characteristics, user's activities/behavior/routine, and services.
- Adaptivity: adaptivity should enhance significantly the security and trustworthiness of the middleware and of the large number of independent, autonomous, heterogeneous and interacting sub-systems by incorporating novel technologies that promote their autonomous/autonomic management when addressing attacks and operational failures. The system should be able to recognize unmet needs within its execution context and to adapt itself to meet those needs.
- Lightweight: minimum range of functionality used by most applications.
- Flexibility: all middleware layers will be easily configurable through an administration API that will be accessed through management consoles. Properties such as the routing, conversion and storage of data, will be capable to be configured at runtime.
- Extensibility: on top of the middleware, it will be possible to easily add new smart services that aggregate on top of the gathered data, as well as to plug data consumers. Both approaches allow generating relevant information on top the integrated data that was collected by the integrated systems.
- Standards-compliance: this project will utilize open standards for interfaces definition, network communication, and data representation, also allowing the extensibility of the middleware by facilitating the integration of additional sub-systems and services.

Figure 1 depicts the conceptual middleware purpose. At the core, it will provide the mechanism to integrate mobile devices, social networks and health sensors; to derive a general architecture enabling general interoperability and is based in the use of an intelligent agent. Figure 2 depicts the proposed middleware architecture. The proposed context-aware system can be represented as a layered middleware composed from bottom to top by sensors, raw data retrieval, preprocessing, storage or management, and an application layer. This approach will allow for the identification of common concepts in both context-aware and prediction computing frameworks, allowing us to devise a general concept for smarter device development. It will be possible to pool data from smart devices in terms of context awareness: for text mining, sentiment analysis, node classification in the context of this application domain. Individual users will be able to automatically convert "units" from smart devices and export and send data/reports to the physicians, health groups, hospitals and even social networks. In short, the

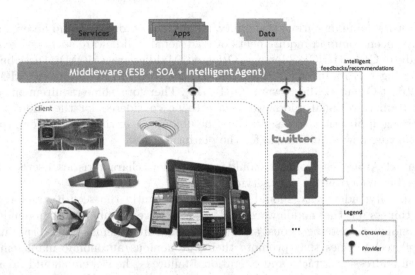

Fig. 1. The high-level view of the Device Nimbus concept

proposed middleware aims to collect and process data from multiple sources, as well as to infer events or patterns that suggest more complicated circumstances.

Data Collectors. The Data Collectors are the part of the middleware that is responsible for collecting data from different devices and sensors. The objective of this layer is to allow for the easy integration of sub-systems that collect data from the external world. This data collection can be achieved via sensors that provide data, through people that feed the system with data (e.g. through mobile devices), or through systems that are able to gather non-structured data from the Web (e.g. social networks, web pages, documents - intelligent agents should be used as consumers of Third-party applications - private protocols). A key aspect is data from this layer may come from different domains (e.g. fitness, illness, weather), which ultimately will allow the Data Integration Layer to extract cutting-edge information for supporting more advance smart services. Due to the dynamic nature of sensors/systems that may enter or leave the middleware in an unpredictable way, we decided to use a dynamic services platform in order to bring SOC to this layer. Both dynamicity and flexibility that allow the evolution of components and services at runtime, among other reasons, made the OSGi the platform of choice for constructing this layer. Built on top, an ESB is responsible for receiving data from different sensors and systems and delivering them into the middleware.

Data Integration. The second key block in the design, is the part of the middleware that is responsible for persistence and data integration. The data collected from the Data Collection Systems and persisted in an environment that relies e.g. on a Cloud Computing infrastructure for guaranteeing the provisioning of

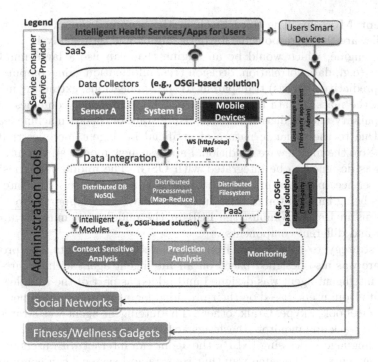

Fig. 2. The high-level view of the Device Nimbus architecture

the necessary storage space. Moreover, the middleware is able to handle several communication protocols thanks to bridge mechanisms we provide: Java message service (JMS) and web services (HTTP/SOAP). Bridges to other protocols, such as XMPP, could be easily added even during execution. The ESB and the intelligent agent of the middleware manage the input data from the different devices in a consolidated NOSQL database.

A significant challenge when developing smart applications, as well as interoperability of distributed systems, is the design of techniques for the integration of distributed data on the Web and from sensors. Processing and analysing acquired data, associated with concepts of pervasive and ubiquitous computing, among others, supports smart applications and context-sensitive systems development. The proposed data integration module is developed for ensure big data processing, classification and organization to support the development of applications on top. Additionally, a service layer providing access to the processed data allows the construction of smart applications and services that reuse functionality of the platform.

The data integration module can be developed with third party components and engines for processing the data and inferring information and knowledge from it. Mechanisms for data analysis and data mining must be used in this module.

Intelligent Module. The third block in the proposed design is the part of the middleware that is responsible for data mining and data analysis. The data-reasoning engine, which would be an engine that can use a large number of techniques (e.g., data correlation, decision tree, information gain, computational context, predictive analysis) transparent to the applications, extracting information from the massive data that are stored. This component combines context aware capability and predictive analysis, using state-of the-art machine learning models. Due to the dynamic nature of artificial intelligent modules that may enter or leave the middleware (additional analysis new modules), the intelligent module was designed to be integrated with OSGi. The middleware considered dynamic services platform in order to bring SOC to this layer. The Intelligent Module is being built on top of Data Integration layer. To support more sensors and systems in Device Nimbus, the ESB in the middleware must be updated with new and different components.

As a strategy to collect data from more web environments, the proposed middleware was also designed based on an intelligent agent architecture-based use. The intelligent agent was designed and added as part of the Middleware to collect data from users, based on their provided logins in web environments (e.g. Twitter, Facebook, Skype, Gtalk, other). The intelligent agent was strategically designed to track and monitor #hashtags and, to work as a chatterbot. Through Natural Language Processing (NLP), the agent can interact with users in different environments. In the same way that sensors and systems can communicate with Device Nimbus to provide data (through the ESB), users can provide data (logins in social networks, devices that they want to be tracked and monitored, other), to the middleware through simple and natural chats with the intelligent agent.

To better explain how the intelligent agent works in the middleware, we present an example scenario that describes the interaction between one user and the Device Nimbus. In this scenario, we assume that the middleware will be able to collect data from heterogeneous and distributed **Sensors (S)**, considering **Context Elements (CE)** to answer **Questions (Q)**:

- **S** = {Humidity, Temperature, NFC, Luminosity, Facebook, Twitter};
- **C** = {{New posts in Facebook and Twitter, from the middleware users', using nikeplus or runkeeper apps},{Interactions between users' and the intelligent agent of the middleware through Facebook or Twitter - NLP},{Big climatic changes},{Holidays and special dates}};
- **Q** = {{Identify runners in a specific location, based on data collected from Twitter or Facebook (#hashtag) posts}; {Identify the relationship between runners in a specific location and environmental data (temperature/humidity/date/time)}; {Identify whether the same runner visited different locations using Facebook or Twitter data}; {Identify the main running locations in the city (city mapping)}; {Identify the main running locations in the city and what are the most empty/crowd date/time}; {Identify the main running locations in the city, what are the most empty/crowd date/time and the relationship between these locations and environmental data (temperature/humidity)}; {Identify

what individuals has to say about the running locations of the city by tracking Twitter and Facebook hashtags}.

To answer the Questions (Q), the Intelligent Agent must be able to merge inputs from the distributed and heterogenous sensors (S), considering the different CE and routine of each user. There are people that like running in the rain, while others love running in sunny days, for example. To identify the main fitness locations in the city, the Intelligent Agent must be able to track the users that decide to be tracked by the middleware. By merging their fitness location, day of the week, time of the day and frequency that they run/exercise, the intelligent will be able to provide rich and personalised feedbacks to each user, based on their needs. If a user loves to run with friends, maybe it's better to go to a crowded location instead of trying to meet people in an empty location.

To ensure the quality of data collected from Twitter and Facebook, the intelligent agent and the ESB are both looking for the same #hashtags and users (logins were provided as input). A single instance of an intelligent agent, which is provided by Device Nimbus middleware, can be available in lots of different environments (such as a contact on Skype and GTalk or as an user in Twitter or Facebook). Despite the fact that the intelligent agent tracks special #hashtags from Device Nimbus users and appears in many different environments, the middleware provides a single agent to them all, which. In other words, the user can chat about his health or routine across different environments with the same bot. If a user starts communicating with the intelligent agent in GTalk, asking him about good spots to run: "where can I run in Melbourne?" he will get an answer about it, as requested. In parallel, the intelligent agent will be monitoring lots of different users on Twitter and Facebook, and will be able to identify where most of them are running in the city, days of the week that people most run, time of the day, and others. To provide best answers, the data collected from environmental sensors and other data sources will be also considered. By providing an interface of Device Nimbus intelligent agent as a chatterbot, more data can be collected and analysed from different users in different environments.

The concept of an intelligent agent monitoring users in different environments was presented in [13,14], and was here adapted to the fitness and wellness domain. To be tracked/monitored by Device Nimbus, each user can just use special commands such as "#addEnvironment Twitter oliveiraeduardo" to set a new login in the middleware (user is in Facebook adding a Twitter login, for example - teaching the bot his others logins distributed in the Web). The advantage to share logins with Device Nimbus is because the middleware with provide health support and assistance to the user. Only the owner of each login recorded in the middleware have access to their personal information and feedback.

For the Natural Language Processing, used by the intelligent agent of the middleware to communicate with the users in the various integrated Web environments, we used the ProgramD library and Drools inference engine (rule-based reasoning). Drools is responsible for integrating users distributed data and for considering context while users are interacting with the intelligent agent of Device Nimbus. The knowledge-api, drools-core, drools-compiler and drools-decisiontables modules are working with OSGi. ProgramD is a fully functional Artificial Intelligence Markup Language (AIML) bot engine that is implemented with Java. It supports multiple bots, it is easy to configure and runs in a GUI application and also under a J2EE environment. AIML is an XML dialect for creating natural language software agents. When the AIML markup language is loaded for the bot, users can chat with the bot.

The advantage of providing a single intelligent agent in the middleware lies in the fact that with only one agent, Device Nimbus can also have a single integrated database. If a user interacts with the intelligent agent through Facebook, the agent will know, referring to the historical database of the user that he has already communicated with him through Twitter and Skype, and that s/he has demonstrated interest in running spots. At the same time, the intelligent agent is able to integrate these data with data that comes from #hashtags or other different wearable devices, modelling every user based on their routine and unique needs. Device Nimbus middleware provides also an interface to help in configuring, monitoring and managing the Middleware and to get connected with Third-Party apps, as described below:

Administration Tools. The final components of the model that must be addressed are administration tools. The proposed middleware provides an array of administration tools that allow users configuring, monitoring and managing of the subsystems Fig. 2. This basically includes (i) a system management console for visualizing the nodes that participate in the system's architecture (which may vary over time) at runtime, and eventually reconfiguring system parameters on them; and (ii) a tool for visualizing and configuring the system monitoring and adaptation policies. The goal is to provide a sort of administration view (i.e. a control panel) for the people that will be in charge of the system administration.

Intelligent Healthcare Services. User interaction with the middleware is via intelligent health services/apps. Applications based on service-oriented computing will benefit from the middleware, which will provide many services on top of the data that is pre-processed Fig. 2. Examples of such applications are a command and control centre that visualizes the data and analytic information about what is being collected from the data collection systems, and an application that shows trends/predictions about health domain.

In summary, Device Nimbus is designed in order to achieve three major measurable objectives: (1) definition and implementation of components for the data collection, (2) definition and implementation of components for the data integration, (3) definition and implementation of a layer for processing and analyzing the acquired data.

In order to test the proposed middleware and validate the three main components of Device Nimbus, a minimum viable product (MVP) is under development and will be detailed, with results, in the future. A series of tests is being strategically planned to measure the efficacy of the MVP implementation of Device Nimbus. Given constraints, we are nominating fitness and wellbeing apps as our primary source of data from the wider health domain. As a second step, experiments will be conducted using some of the health apps listed in Sect. 3 instead of the fitness and wellbeing apps. Such experiments will require collaboration with medical professionals, clinics and hospitals.

Each of the components of the proposed middleware will tested sequentially:

- Collecting heterogeneous data: (i) Test data collection from Twitter, Nikeplus, RunKeeper, Gtalk and Skype; (ii) Test data collection from environmental sensors (temperature, humidity, noise and luminosity).
- Integrating data: (i) Test the ability to the middleware to integrate the heterogeneous data into the NoSQL database; (ii) Test the ability of the intelligent agent of the middleware to integrate the users data into the NoSQL database.
- Analyzing data: (i) Test the ability to the middleware to analyse the integrated data (Context Sensitive Analysis).

By collecting, integrating and analysing data, the proposed middleware will be able to answer the Questions (Q) presented before in this Section.

3 Related Work

Many of the existing ICT solutions for smart health device are proprietary, usually provided by large services vendors, e.g. the likes of IBM, Microsoft, Google, Samsung, Apple and others [2]. These solutions/products are designed as a unified, distributed and real-time control platform, adding cloud computing, sensing, simulation, analysis services and applications. Integrated sensor networks, mobile devices and people power these systems, which are able to combine, aggregate, analyse and inspect for deriving knowledge from health settings. Current developments focus on wearable technologies like smart watches instrumented to collect health data, such as: *physiological sensors* to collect heart rate, blood pressure, respiration rate, electrocardiogram, and others; *environmental sensors* that collect external temperature, velocity, acceleration, and others, and; *light reflection sensors* to collect health parameters like oxygen saturation, skin temperature, blood pressure, and others. These efforts supersede early works based on designed instrumentation, such as [3,16], and offer commodity solutions for experimentation with advanced health monitoring.

The combination of sensors, devices and systems from different modalities and standards makes it necessary to develop hardware independent software solutions for efficient application development [9,17]. In this context, there are numerous studies focusing on middleware/platform design [5,10,19]. Middleware can help health sensor/mobile networks to manage their inherent complexity and heterogeneity. The idea is to isolate commons behaviour that can be reused by several applications and to encapsulate it as system services [1].

As a way of avoiding proprietary solutions, the Open Health Tools was created as an open source community with a vision of enabling an ecosystem, where members of Health and IT professions collaborate. This collaboration is based on building interoperable systems (platform) that enable patients and their care providers to have access to vital and reliable information at the time and place it is needed. However, interoperability benefits are highly dispersed across many

Table 1. State-of-the-art comparison

References x requirements	Context	Adaptivity	Lightweight	Flexibility	Extensibility	Open softwares
Microsoft HealthVault	X	X	X	X		
Samsung S Health	X	X	X	X		
Apple iHealth	X	X	X	X		
Open Health Tools		X	X			X
Magic Broker [4]		X	X	X	X	
SOFIA [6]		X	X	X	X	
Xively		X	X	X	X	
Device Nimbus	X	X	X	X	X	X

stakeholders, and early adopters are penalised by negative network externalities and first-mover disadvantages, e.g. faced barriers and challenges that have resulted in partial success, slow progress and outright failure [12].

Although other initiatives like Magic Broker [4], SOFIA [6] and Xively provide part of their infrastructure as free and open source software, they describe system architectures that are just centred on the smart environments concept, for creating interacting objects ecosystems (sensors, devices, appliances and embedded systems). They are not concerned with simultaneously dealing with different vertical axes (e.g., health, weather, fitness and feeding) and with closed or proprietary wearable solutions. In addition, all of those experimental efforts are custom-tailored solutions whose components or subsystems were not modelled as interchangeable parts, nor were conceived to be integrated with other subsystems. In health domain, a big challenge is to collect, integrate and analyse data from proprietary solutions (mobile devices from Nike, Garmin, Google, Apple and others) and from lots of proprietary different sensors and, tracking the same user based in lots of different input devices to provide them contextual feedbacks/suggestions, based on their particular needs. Based on the requirement analysis presented in Sect. 2, we position "Device Nimbus" against the state-of-the-art as described in Table 1. It becomes evident that the proposal covers a gap in current technology not previously contemplated by the analysed technology.

4   Conclusion

We propose an end-to-end solution for continuous health monitoring enhanced with support to contextualised recommendation. The proposal combines environmental monitoring, personal data collecting, and predictive analytics. For that, we introduce a middleware called Device Nimbus. It provides the structures to integrate data from sensors in existing mobile computing technology. Moreover, it includes the algorithms for context inference and recommendation support. An important design feature of the middleware is the inclusion of an active intelligent agent and an environment for the integration and aggregation of services, applications and fitness/wellness data. The rationale, on which our middleware design is based, is that it should be flexible and extensible.

We demonstrate how the propose improves the state-of-the-art by supporting the requirements for Context-Awareness, Adaptivity, Lightweight, Flexibility, Extensibility and Standards-compliance. The middleware encapsulated key blocks (or components) for data collection from heterogeneous sensors and devices; data retrieval; pre-processing; storage or management, and an application layer. The intelligent agent module allows for prediction capabilities and context awareness, allowing for the dissemination of useful and timely information for users in dynamic environments. As future work, we intend to develop and test the MVP in the fitness and wellness domain. Further implementation and partnership with clinics and hospitals will allow Device Nimbus to provide users' with personalized preventive care, providing early warning signs for serious illnesses, by collecting physiological inputs such as heart rate, blood pressure, body and skin temperature, oxygen saturation, respiration rate, electrocardiogram and others.

The middleware will also provide an environment for the integration and aggregation of services and applications in the health domain. The Device Nimbus platform also aims to develop a business ecosystem that targets Health related domains, enabling the participation of different solution providers and thus stimulating the economy in the ICT sector. Third party services will be able to built systems and apps on top of the platform thus forming an ecosystem around the platform. This proposed work is currently under development in a pilot project in The University of Melbourne - Australia, in collaboration with the University of Campinas and the CESAR.

References

1. Alemdar, H., Ersoy, C.: Wireless sensor networks for healthcare: a survey. Comput. Netw. **54**(15), 2688–2710 (2010)
2. Oliveira, E.A., Kirley, M., Vanz, E., Gama, K.: HSPY: an intelligent framework for context and predictive analysis for smarter health devices. In: 2014 International Conference on Information and Communication Technology Convergence (ICTC), pp. 53–58. IEEE (2014)
3. Black, J., Koch, F., Sonenberg, L., Scheepers, R., Khandoker, A., Charry, E., Walker, B., Soe, N.L.: Mobile solutions for front-line health workers in developing countries. In: 2009 11th International Conference on e-Health Networking, Applications and Services (Healthcom), pp. 89–93 (2009)
4. Blackstock, M., Kaviani, N., Lea, R., Friday, A.: Magic broker 2: an open and extensible platform for the internet of things. In: Internet of Things (IOT), pp. 1–8. IEEE (2010)
5. Chatzigiannakis, I., Mylonas, G., Nikoletseas, S.: 50 ways to build your application: a survey of middleware and systems for wireless sensor networks. In: IEEE Conference on Emerging Technologies and Factory Automation, 2007 ETFA, pp. 466–473. IEEE (2007)
6. Filipponi, L., Vitaletti, A., Landi, G., Memeo, V., Laura, G., Pucci, P.: Smart city: an event driven architecture for monitoring public spaces with heterogeneous sensors. In: 2010 Fourth International Conference on Sensor Technologies and Applications (SENSORCOMM), pp. 281–286. IEEE (2010)

7. Gatzoulis, L., Iakovidis, I.: Wearable and portable ehealth systems. IEEE Eng. Med. Biol. Mag. **26**(5), 51–56 (2007)
8. Koster, A., Koch, F., Kim, Y.B.: Serendipitous recommendation based on big context. In: Bazzan, A.L.C., Pichara, K. (eds.) IBERAMIA 2014. LNCS, vol. 8864, pp. 319–330. Springer, Heidelberg (2014)
9. López, T.S., Kim, D.: Wireless sensor networks and rfid integration for context aware services. Artigo publicado no Web Site do Auto-ID Labs (2008). http://www.autoidlabs.org/uploads/media/withhold_AUTOIDLABS-WP-SWNET-026.pdf
10. Lu, H.F., Chen, J.L.: Design of middleware for tele-homecare systems. Wireless Commun. Mob. Comput. **9**(12), 1553–1564 (2009)
11. Lymberis, A., Dittmar, A.: Advanced wearable health systems and applications-research and development efforts in the European union. IEEE Eng. Med. Biol. Mag. **26**(3), 29–33 (2007)
12. Manzaroli, D., Roffia, L., Cinotti, T.S., Azzoni, P., Ovaska, E., Nannini, V., Mattarozzi, S.: Smart-m3 and OSGI: the interoperability platform. In: 2010 IEEE Symposium on Computers and Communications (ISCC), pp. 1053–1058. IEEE (2010)
13. Oliveira, E.A., Tedesco, P.: i-collaboration 3.0: um framework de apoio ao desenvolvimento de ambientes distribuídos de aprendizagem sensíveis ao contexto. In: Anais dos Workshops do Congresso Brasileiro de Informática na Educação, vol. 2 (2013)
14. Oliveira, E.A., Tedesco, P.: i-collaboration: um modelo de colaboração inteligente personalizada para ambientes de ead. Revista Brasileira de Informática na Educação **18**(1), 17–31 (2010)
15. Park, N.S., Lee, K.W., Kim, H.: A middleware for supporting context-aware services in mobile and ubiquitous environment. In: 2005 International Conference on Mobile Business, ICMB 2005, pp. 694–697. IEEE (2005)
16. Rolim, C., Koch, F., Black, J., Geyer, C.: Health solutions using low cost mobile phones and smart spaces for the continuous monitoring and remote diagnostics of chronic diseases. In: The Third International Conference on eHealth, Telemedicine, and Social Medicine, eTELEMED 2011, pp. 72–76 (2011)
17. Triantafyllidis, A., Koutkias, V., Chouvarda, I., Maglaveras, N.: an open and reconfigurable wireless sensor network for pervasive health monitoring. In: 2008 Second International Conference on Pervasive Computing Technologies for Healthcare, PervasiveHealth 2008, pp. 112–115. IEEE (2008)
18. Tröster, G.: The agenda of wearable healthcare. In: IMIA Yearbook of Medical Informatics, pp. 125–138 (2005)
19. Waluyo, A.B., Ying, S., Pek, I., Wu, J.K.: Middleware for wireless medical body area network. In: 2007 IEEE Biomedical Circuits and Systems Conference, BIOCAS 2007, pp. 183–186. IEEE (2007)

The Influence of Users' Personality on the Perception of Intelligent Virtual Agents' Personality and the Trust Within a Collaborative Context

Nader Hanna[✉] and Deborah Richards

Department of Computing, Macquarie University, Sydney, NSW 2109, Australia
{nader.hanna, deborah.richards}@mq.edu.au

Abstract. As Intelligent Virtual Agents (IVAs) have been widely used for applications that require human interaction and collaboration, modeling an IVA that can exhibit personalities is becoming increasingly important. A large body of research has studied variant verbal and non-verbal aspects that are used to deduce an IVA's personality; however, research falls short in showing whether humans' personality influences their perception of the IVA's personality. This paper presents an empirical study that investigated whether human users can perceive the intended personality of an IVA through verbal and/or non-verbal communication, on one hand, and the influence of the user's own personality on their perception, on the other hand. Furthermore, we investigated whether the perceived personality had an impact on the human's level of trust in the IVA teammate. The results showed that similarity in personalities between humans and IVAs tended to significantly influence the humans' correct perception of the IVA's personality and that different perceived personalities influenced the human's level of trust.

Keywords: Personality traits · FFM · Extraversion · Agreeableness · Intelligent Virtual Agent · Human-IVA collaboration · Trust

1 Introduction

An Intelligent Virtual Agent (IVA) is a term used to define an autonomous entity in a virtual environment. This entity should not only look like, but also behave as a living organism (e.g., human, animal, imaginary creature) [1]. Several studies aimed to create believable IVAs and include sophisticated characteristics similar to humans. Among these characteristics, researchers have sought to create unique IVAs with distinct personalities. Personality is a personal aspect that makes it possible to distinguish between different people [2]. Because our personality affects our internal perception and actual behavior [3], personality has been included in multiple aspects of IVAs including their expressive aspect, i.e. non-verbal communication and verbal communication, and their internal aspect, such as planning [4].

A number of psychological theories proposed foundations to understand personality, yet one of the most well-known and widely-accepted theories is Five-Factor Model (FFM) of personality [5]. FFM model is comprised of five-personality dimensions:

© Springer International Publishing Switzerland 2015
F. Koch et al. (Eds.): CARE-MFSC 2015, CCIS 541, pp. 31–47, 2015.
DOI: 10.1007/978-3-319-24804-2_3

openness to experience; conscientiousness; extraversion (antonym-introversion); agreeableness (antonym antagonism); and neuroticism. After its wide success in understanding humans' personalities, numerous studies used the foundations of FFM to personalize the behaviour of IVAs [6]. IVAs with personality according to FFM have been studied in different contexts including interviews, medical treatment, and interactive narrative [7].

Many research papers used FFM to influence the multi-modal behaviour of IVAs and the aim was to investigate whether humans can predict IVAs' personalities based on their expressed behaviour (e.g., [8]), however these papers used a simple simulated environment and basic human-IVA interaction. IVAs are increasingly being used as teammates in heterogeneous teams that combine both IVAs and humans. Studying IVA's personalities in the context of teamwork with humans has become a recent interesting topic to study [7]. A few studies go beyond basic interaction and show more complex scenarios such as interaction in a collaborative environment. Collaborative environments require both human users and IVAs to work as team members to achieve a shared task. Collaborative situations make humans' prediction of IVAs' personalities more difficult as humans tend to focus on achievement of the task.

Among these few studies that investigated IVA's personality in a collaborative context, Negrón et al. [9] stressed the integration of nonverbal communication cues in IVAs as a way to provide human team members with alternatives that may accelerate the communication process and foster collaboration. However, in that work IVAs played the role of facilitator to the human team and were not real teammates. In another study, Prabhala and Gallimore [10] found that people could perceive personality from avatars through their actions, language, and behavior. However, this research falls short in showing whether humans' personality influences their perception of the IVAs' personality.

A further consideration in our study, and gap in existing work, concerned the influence of the IVA's perceived personality on the human's level of trust in the IVA. Trust is widely recognized as an important facilitator of successful relationships and essential in the context of successful collaboration [11]. Trust has been defined as an individual's belief in another person's capabilities and honesty based on his/her own direct experiences [12]. A commonly used classic definition of trust is *"the willingness of a party to be vulnerable to the actions of another party based on the expectation that the other will perform a particular action important to the trustor, irrespective of the ability to monitor or control that other party"* (p. 712) [13]. This definition considers trust as a transient state in any particular situation. Many other definitions of trust exist that offer alternative perspectives (see [14]); nevertheless, the majority of these definitions share the concept of expectation and confidence in the other actors' reliability, fairness and integrity.

The paper is organized as follows. Section 2 presents related work on IVAs and personality. Section 3 presents further background about the FFM personality model and how it has been incorporated in the design of our IVA's verbal and non-verbal behaviour. Section 4 presents the research questions. The methodology used to answer the research questions is presented in Sect. 5. The results will be given in Sect. 6, followed by discussion in Sect. 7. Finally, Sect. 8 presents conclusions and future work.

2 Literature Review

Many researchers have been working on human-IVA relationships [15–18]. Numerous studies have considered whether human participants are able to perceive IVA's personality through communication with IVA. Doce et al. [4] presented a model to create an IVA with distinguishable FFM personality traits. In their model, four cognitive/behavioural processes were identified that were strongly influenced by personality: emotions, coping behaviour, planning and bodily expression. The personality traits were adopted to influence each of these processes. Users were asked to classify different personalities for IVA. Although users' classification correlated with the original values for extraversion, neuroticism, and agreeableness, users failed to identify conscientiousness. Moreover, the model did not introduce personality in IVA's verbal communication.

Rushforth et al. [19] presented an initial attempt to build a personality framework for virtual characters that allows the domain designer to author different personalities for the same character. The results of two experiments showed that the presented framework had an impact on user perception of several aspects of the personality of the virtual character. Neff et al. exploited the extraversion [20] and neuroticism [21] traits of the Big Five model in multimodal characters evaluating the effects of verbal and nonverbal behavior in personality perception studies. Cafaro et al. [22] conducted a study to investigate how IVA's non-verbal communication influence the first encounters between humans and virtual agents. Each agent exclusively exhibited nonverbal cues (smile, gaze and proximity), and then participants judged IVA's personality (extraversion) and interpersonal attitude (hostility/friendliness) based on the nonverbal cues. The results showed that participants could form an impression about the IVA's personality from the observed non-verbal behaviour. Kang et al. [23] explored associations between the five-factor personality traits of human subjects and their feelings of rapport when they interacted with a virtual agent or real humans.

Despite the large body of research in human perception of IVAs' personality, little research considers personality in a collaborative context. Among these few studies, Aguilar et al. [24] propose a Team Training Strategy whose purpose is to promote social skills. In this training strategy, personality traits have been assigned to appropriate team tasks. However, their study did not investigate the interaction between the personalities of both humans and IVAs.

3 Intelligent Personality Traits: Five-Factor Model

In the last 50 years, the FFM model of personality has become a standard in the field of classifying personalities. FFM [25] claims that personality varies on five factors: Openness, Conscientiousness, Extraversion, Agreeableness and Neuroticism. Openness means being open to experience new things, being imaginative, and intelligent. Conscientiousness indicates responsibility, reliability and tidiness. Extravert personality is outgoing, sociable, assertive and energetic. Agreeableness means a person is trustworthy, kind and cooperative by considering others' goals. A neurotic character is anxious, nervous and prone to depression and lacks emotional stability.

Studies that have explored personality traits and teamwork stress the role of both extraversion and agreeableness to foster inter-relationships between team members. Extraversion and agreeableness were selected in our study because they have been shown to be predominant traits in collaboration and teamwork [26]. The extraversion trait affects interpersonal relations through the quality of social interactions [27, 28]. Extraverts are usually active members in teamwork interactions and often popular among their mates [29].

3.1 Expressing Personality Through Verbal Behaviour

Our personality is likely to influence how we speak [30]. Speaking style can reveal certain personality traits; some traits are easier to detect than others [31]. A number of studies have used verbal capabilities to represent different IVA personalities [32]. Neff et al. [20] determined a number of aspects that demonstrate the impact of the IVA's extravert personality on the IVA's verbal behaviour. Among the list of aspects mentioned in [20], we selected the dominant aspects, see Table 1, as the basis of the design of the IVA in our study. Verbal messages were designed and reviewed by the authors. The messages were designed according to the criteria in Table 1.

Table 1. Verbal aspects used to express introversion/extraversion in IVA's behaviour

Parameter	Description	Introvert	Extravert
Verbosity	Control the number of propositions in the utterance	Low	High
Restatements	Paraphrase an existing proposition	Low	High
Request confirmation	Begin the utterance with a confirmation of the propositions	Low	High
Emphasizer hedges	Insert syntactic elements (really, basically, actually, just) to strengthen a proposition	Low	High
Negation	Negate a verb by replacing its modifier by its antonym	High	Low
Filled pauses	Insert syntactic elements expressing hesitancy	High	Low

3.2 Expressing Personality Through Non-verbal Behaviour

A number of studies addressed how the extraversion personality trait can be represented in an IVA's non-verbal signaling. As verbal behaviours have already been identified that show an IVA's personality, Doce et al. [4] proposed several non-verbal features that could be used to show personality traits in IVA, these features include:

- Spatial extent – the required amount of space to perform an expression - extraverts use a lot of spatial extent, while introverts use a small space.
- Temporal extent - amount of time spent to perform an expression - we assigned a short temporal extent to extraverts.

- Fluidity - smoothness of movements - agents have a high fluidity if they are not extraverted nor neurotic and a low fluidity otherwise.
- Power – intensity of an intention - power is directly proportional to extraversion.
- Repetitiveness -repetition of certain movements - a character with high extraversion will have high repetitivity.

Additionally, the IVA's physical position relative to the human's view or their avatar has been investigated. Argyle's [33] status and affiliation model for animating non-verbal behavior of virtual agents identified two fundamental dimensions for non-verbal behavior: affiliation and status. Affiliation can be characterized as wanting a close relationship and it is associated with non-verbal clues such as close physical position. Other studies suggested that agents approaching the subject's avatar were judged as more extraverted than agents not approaching them, regardless of smile and the amount of gaze they gave [22]. In the design of our agent, we chose the dominant features, shown Table 2.

Table 2. Non-verbal aspects used to express introversion/extraversion in IVA's behaviour

Parameter	Description	Introvert	Extravert
Spatial extent	Amount of space required to perform an expression	Low	High
Temporal extent	Amount of time spent to perform an expression	Long	Short
Repetitivity	Repetition of certain movements	Low	High
Body position	Close physical postures	Far	Close

4 Research Questions

We have proposed the following research questions to investigate the influence of IVAs' personalities on their verbal and non-verbal communication as perceived by humans; in addition, we explored the relation between (mis)match in human-IVA personalities and humans' right perception of IVAs' personality. Finally, we consider how IVA personality and the match with humans was linked with human trust in IVA's decisions and recommendations. Figure 1 shows an overview of the research model that underpins the following research questions:

Q1: Can IVAs' multimodal communication, i.e. verbal and non-verbal communication, distinguish the IVA's extravert/introvert personality?
Q2: Can IVAs' multimodal communication distinguish their agreeableness/antagonism personality?
Q3: Does a match in the human and IVA's extravert/introvert personality traits influence the human's perception of the IVA's extravert-introvert personality trait?
Q4: Does a match in the humans and IVA's agreeableness/antagonism personality traits influence the human's perception of the IVA's agreeableness/antagonism personality trait?
Q5: Does perceiving the IVA's agreeableness/antagonism or extravert/introvert personality traits influence the human's trust in the IVA?

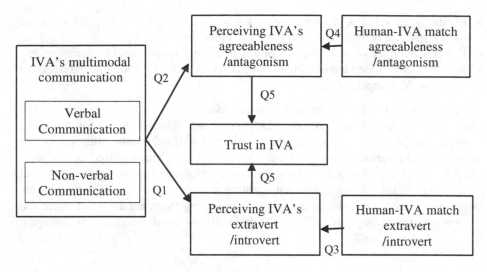

Fig. 1. The proposed research model

5 Materials and Method

An experiment was conducted to answer the five research questions. The design, the participants, the procedure, and the scenario are described below.

5.1 Experimental Design and Procedure

The study was structured as 2×2 between-subject experiment and a control group.- Each subject had to take just one treatment. The experiment consisted of five different treatments with the same virtual scenario but the IVA had different personalities, see Table 3. One treatment was a control with a neutral personality IVA. The other four experimental treatments had the four combinations of the two studied personality traits, i.e. extraversion and agreeableness. The four combinations were extraversion-agreeableness, extraversion-antagonism, introversion-agreeableness and introversion-antagonism. Participants had to access a web-based system that contained the five treatments and managed treatment assignment. Each participant was assigned one of the five treatments. The assignment was done by the system sequentially and equally. Participants were divided into five groups each containing 11 students. Participants used the virtual system individually so that the collaboration would be one-to-one between him/herself and the agent. We dedicated 20 min for the study that consisted of four parts, as below, in one session.

- Part 1: sign consent forms and complete biographical information.
- Part 2: Complete 7-item personality test to measure the two personality traits using International Personality Item Pool (IPIP) [34].
- Part 3: Participation in the scenario in the 3D virtual scene. In the beginning of the scenario, the participants were provided with online instructions about the goal of

Table 3. The level of extraversion and agreeableness personality traits in each treatment

Treatment	Extraversion	Agreeableness
Extraversion-agreeableness	High	High
Extraversion-antagonism	High	Low
Introversion-agreeableness	Low	High
Introversion-antagonism	Low	Low
Neutral	Averaged	Averaged

the virtual scenario, the name and the use of each tool in the toolbox and how to select/close the verbal messages.

- Part 4: Complete 5-item survey (5 items each for verbal and non-verbal communication and 5 items for measuring trust) that measures the participant's perception of the communication and collaboration experience. Additionally participants completed a test of the perceived personality of the IVA by answering four items of the Ten-Item Personality Inventory (TIPI) [35].

Both personality tests, i.e. IPIP and TIPI used a 5-item Likert Scale, where 1 corresponded to "Disagree Strongly" and 5 to "Agree Strongly". Additionally, all inputs from the user were logged to allow recreation of navigation paths and record inputs such as responses and selected tools. These inputs included selected regions in the scenario. Analysis of interaction logs to find the most frequently triggered stimuli in the scenario was used before in other studies [32].

5.2 Participants

Fifty-five (55) second-year undergraduate students enrolled in a science unit completed the collaborative task. Participants were aged between 18 and 51 years (mean = 22.56; SD = 6.95). English native speakers were 94.55 % of the participants. The non-native English speakers had been speaking English on a daily basis on average for 13 years. On a scale of 6 levels (level 1 the least experienced and level 6 the highest experience), 23 % of the participants described themselves as having basic computers skills (level 2), 5.45 % as having advanced skills (level 6), while 70.91 % said they have proficient computer skills (level 5). To measure game and 3D application experience, participants answered the question "How many hours a week do you play computer games?" with responses ranging between 0 to 25 h weekly (mean = 2.73, SD = 4.69).

5.3 Case Study

In order to answer the proposed research questions a collaborative scenario need to be designed. This scenario needs to include the common features encompassed in any collaborative scenario. These features are as follows.

The Features of Collaborative Scenarios. A number of attempts have been made to define the elements of collaborative activity. In a series of studies, Traum et al. [36] identified the features of collaborative tasks that serve to test out the development of a shared understanding:

– Sharing of the basic facts about the task…sharing the beliefs about the task between collaborators. Traum et al. [36] stressed that it is important to share the basic information not only in an indirect way such as using a whiteboard but also in an intrusive ways such as via dialogues or invitation to perform actions.
– Interferences about the task… the requirement is directly connected to the goal of the collaborative task. The inferences are explicitly negotiated through verbal discussion.
– Problem-solving strategy…As the collaborative activity includes a task to achieve; partners need to have a strategy to accomplish this task. This strategy is individual to each team member, but additionally it should take into account a role to the other partner.
– Sharing information about positions…this element is related to sharing information about the position and progress of each party while achieving the collaborative task. The current position of the partner could be deduced through the partners action, while his/her future position could be communicated though discussion.
– Knowledge representation codes…it is important to use clear notations that represent the required knowledge in the collaborative task. For example, using red label to demonstrate crucial or critical knowledge.
– Interaction rules…the rules the partners agree on to manage the interactions while achieving the task.

In line with these requirements, we proposed a scenario where a human and an IVA should collaborate to achieve a shared task.

The Proposed Collaborative Scenario. The aim of the scenario. In the scenario-based activity, the human and the agent (a virtual scientist called Charlie) needed to pass a sequence of four obstacles.

Aspects should be considered. There are a number of aspects that should be considered to design a scenario to test out the proposed research question including the following.

First, the actions of both humans and IVAs must be dependent or interleaved; that is to say, none of them can do the task alone and the contribution of the other teammate is crucial for the success of the task.

Second, the task should be divided into stages or sequences in order to observe the progress in team behaviour and performance.

Third, humans must have the option either to conform to the IVA's requests or select a different decision.

Fourth, the verbal and non-verbal communication should be bidirectional, that is the human and agent can send and receive messages.

Finally, communication must be task-oriented. That is not to say that social-oriented communication would not be beneficial, however, that was beyond the scope of this study. The collaborative scenario was implemented using the Unity3D game engine (www.unity3d.com). The scenario included a task where both a human user and an IVA, namely Charlie, have to collaborate to achieve a shared goal. The goal is to pass a sequence of four obstacles to reach their target (scientific laboratory). The four

Fig. 2. Snapshots from the third obstacle (the bush) and IVA personality is high extrovert and high agreeableness

obstacles were brick wall, wooden gate, bush and hill (see Fig. 2). In order to get over each one of these obstacles both the human and IVA have to select a pair of tools from a toolbox that contains 12 tools (pruning shears, bush hook, hammer, chisel, ladder, rope, matchsticks, matchbox, screwdriver, nipper, shovel and mattock). These tools were picked so that each pair of tools would be complementary, i.e. a single tool cannot work without the function of the complementary tool. For example, the chisel needs the hammer and matchstick needs the matchbox. In addition, each obstacle could be passed using different tools. For example, the bush obstacle could be chopped, burnt or climbed. Hence, there should be agreement between the human and the IVA concerning the best way to overcome the obstacle and to select which pair of tools is most suitable for the task. The interaction between human and IVA occurs via two means:

- Verbal communication: through exchanging messages that convey both human's and IVA's requests, examples from the scenario can be found in Table 4.
- Non-verbal communication: through the IVA's, hand gestures to represent different personalities.

Table 4. Examples of IVA's verbal messages along with the level of personality traits that is represented in each message

Trait	Set	Example
Extraversion/Agreeableness	Low	Not a bad idea, I will grab " + tools [human_selected_tool == 5?6:5] + " to help.
	Medium	Good idea, it will save effort and time. I will grab " + tools[human_selected_tool == 7?8:7] + " to help you to climb the gate.
	High	Wow, it is an Excellent idea, I was thinking of climbing the gate too. Hmmm, it is also much faster than breaking or burning that gate. I will grab " + tools[human_selected_tool == 5? 6:5] + " to help you in tying a ladder

(Continued)

Table 4. (*Continued*)

Trait	Set	Example
Extraversion/Antagonism	Low	I was thinking of opening the wooden gate. Don't you agree?
	Medium	It will be hard to break the gate, would you please think of another way to get over this obstacle? For example, what about opening the gate?
	High	Oh, are you kidding? Do you know how long it is going to take to break that wooden gate? I was thinking of opening the gate. Opening the gate will save lots of time and effort. Don't you think so? Don't you agree with me?

6 Results

To answer Q1, the results, as shown in Table 5, showed that there was a significant difference [$F(2, 52) = 15.014$, $p < 0.01$, $\eta2 = 0.366$] between the groups of participants, who had introvert, extravert or neutral IVA, in their perception to different personality of IVA as expressed by the verbal messages of IVA. Furthermore, to understand the difference between these groups (introvert, extravert or neutral), the average evaluation of the IVA's verbal communication of each group was calculated. The results, as can be seen in Fig. 3, showed that the average perception of introvert, extravert and neutral IVAs was 3.66, 4.32 and 3.42 out of 5, respectively.

Table 5. A summary of one-way ANOVA to show difference between participants in perceiving IVA's introversion/extroversion based on IVA's verbal communication

	Sum of squares	df	Mean square	F	Sig.
Between groups	7.599	2	3.800	15.014	0.000
Within groups	13.160	52	0.253		
Total	20.759	54			

Regarding non-verbal communication, Table 6 showed that there was a significant difference $p < 0.01$ [$F(2, 52) = 11.424$, $p < 0.01$, $\eta2 = 0.30$] between the groups of participants, who got introvert, extravert or neutral IVA, in their perception to different personality of IVA due to the non-verbal messages of IVA. Furthermore, to understand the difference between these groups (introvert, extravert or neutral), the average evaluation of the IVA's non-verbal communication was calculated. Figure 3 showed that average perception of introvert, extravert and neutral IVA was 3.72, 4.30 and 3.78 out of 5, respectively.

To answer Q2, the results, see Table 7, showed that there was a significant difference $p < 0.01$ [$F(2, 52) = 6.086$, $p < 0.01$, $\eta^2 = 0.189$] between the groups of participants, who got agreeableness, antagonism or neutral IVA, in their perception to

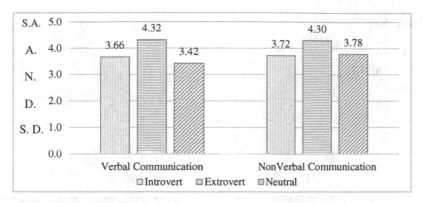

Fig. 3. Average evaluation of verbal and non-verbal communication of Introvert, Extrovert and neutral IVA (S.A. = Strongly Agree, A. = Agree, N. = Neutral, D. = Disagree, S.D. = Strongly Disagree)

Table 6. A summary of one-way ANOVA to show difference between participants in perceiving IVA's introversion/extroversion based on IVA's non-verbal communication

	Sum of squares	df	Mean square	F	Sig.
Between groups	4.178	2	2.089	11.424	0.000
Within groups	9.509	52	0.183		
Total	13.687	54			

IVA's agreeableness-antagonism as represented in verbal messages. To further understand the difference between these groups (agreeableness, antagonism or neutral), the average evaluation to IVA's non-verbal clues of each group was calculated. The results, as can be seen Fig. 4, showed that average perception of agreeableness, antagonism or neutral IVA was 4.15, 3.84 and 3.42 out of 5, respectively. Additionally, the results showed that there was no significant difference between the five groups of participants' in their perception to IVA's agreeableness-antagonism as represented in non-verbal clues.

Table 7. A summary of one-way ANOVA to show difference between participants in perceiving IVA's agreeableness/antagonism based on IVA's verbal communication

	Sum of squares	df	Mean square	F	Sig.
Between groups	3.937	2	1.969	6.086	0.004
Within groups	16.822	52	0.323		
Total	20.759	54			

In answer to Q3, whether the match in extravert/introvert personality trait correlates with humans' perception of IVA's extraversion, the results of Chi-square test, $\chi^2(1, N = 55) = 6.04$, and $p < 0.05$, showed a significant difference between actual match between human and IVA and the correct perception of humans to IVA's extraversion trait.

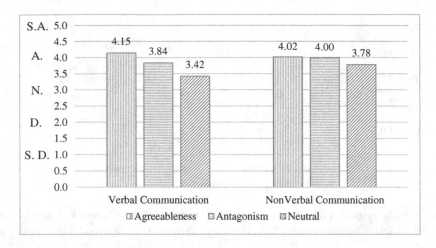

Fig. 4. Average evaluation of verbal and non-verbal communication of Agreeableness, Antagonism and neutral IVA (S.A = Strongly Agree, A. = Agree, N. = Neutral, D. = Disagree, S.D. = Strongly Disagree)

Question four inquired whether the match in agreeableness/antagonism personality trait correlates with humans' perception of IVA's agreeableness. The results of Chi-square test between real match between human and IVA and the correct perception of humans to IVA's agreeableness trait $\chi^2(1, N = 55) = 4.035$, and $p < 0.05$ showed significant difference in the accuracy of the guess of IVA's agreeableness/antagonism personality trait by human users whose agreeableness/antagonism personality match IVA.

The fifth question inquired if the perceived agreeableness/antagonism or extravert/introvert influence humans' trust in IVAs. The results of ANOVA test showed that there was a significant difference between agreeableness/antagonism treatments in human trust in the IVA $p < 0.001$ [$F(1, 53) = 10.93$, $p < 0.001$, $\eta2 = 0.17$]. However, the results of ANOVA test showed that there was no significance difference between extravert/introvert treatments in human trust in the IVA.

7   Discussion

The aim of this paper was to study whether the (mis)match in personality between humans and IVAs influences the humans' perception of the IVAs' personality. To answer this question, we studied the impact of two personality traits, i.e. extraversion and agreeableness, on the perceived multimodal communication, i.e. verbal and non-verbal communication.

The results of the first research question showed that there was a significant difference between the participants in the five treatments in the perception of IVA's extraversion expressed by IVA's verbal and non-verbal behaviour. This finding demonstrated that both verbal and non-verbal communication contribute toward participants' perception of IVA's personality. This result is consistent with the other studies, e.g. [37, 38], that showed the impact of IVA's verbal and non-verbal communication aspects on human users' prediction of IVA's personality.

Regarding the results of perceiving neutral personality, participants did not get different treatments that would allow them to compare between the personalities of IVAs. They were assigned a single treatment and so based on that single experience they perceived the personality of the IVA teammate. Although the neutral personality was meant to be midway between extravert and introvert, participants tended to classify the neutral IVA as either an introvert or extravert. Previous research work has identified the problem of erroneous perception of the neutral emotion and personality, where neutral emotion and personality could be confused with other traits, or other traits could be confused with neutral. In one study, the neutral emotion was easily confused with other emotions such as sadness [39]. In another study, where pictures of an IVA (Alfred) with different head postures and eye gazes were shown, participants were likely to recognize different head posture and eye gaze as neutral [37].

The result of the second research question showed that there was a significant difference between the participants in the five treatments in the perception of IVA's agreeableness as expressed by IVA's verbal behaviour. However, the result did not reveal any significance between participants in differentiating IVA's agreeableness personality because of the non-verbal behaviour of the IVA. The impact of non-verbal behaviour on humans' perception of IVA's personality has been a debated topic. Burgoon [40] suggests that overall approximately 60-65 % of social meaning is derived from nonverbal behaviors. Vinciarelli et al. [41] reported that nonverbal behaviour influences our perception of others. Arellano et al. [37] studied the influence of some visual cues of non-verbal communication, head orientation and eye gaze, on human users' perception of certain IVA personality traits: extraversion, agreeableness and emotional stability. The results showed that non-verbal communication visual clues affected significantly the users' perception of the IVA's personality traits. However, in their study about varying personality in spoken dialogue, Rushforth et al. [19] reported that feedback from the participants suggested that the non-verbal behavior may have been a confounding factor in their perception of IVA's personality.

The results of questions three and four suggested that in the collaborative context the similarity in personality traits between IVAs and humans is likely to impact on humans' perception of IVAs' personality. Numerous studies reported different points of views; while Isbister [42] found people liked virtual agents which showed a different personality to their own, other researchers [43, 44] report that people preferred computer interfaces (including IVA) that embodied a similar type of personality to their own. These differences in findings are probably due to the differences in goals and designs of each of the studies and highlights the complexity of the personality dimension and its effects.

Relating to question five, the results showed that the humans trusted in the agreeable IVA and not the antagonistic or the neutral IVA. Additionally, the results showed that the humans trusted in the extravert IVA rather than introvert or the neutral IVA. Probably agreeableness is the personality trait that can be identified as the most associated with trust. The reason for this strong association is due to the nature of agreeableness that makes the individual willing to conform to the needs of others. Some researchers have claimed that the propensity to trust is a facet or component of agreeableness [45]. Many studies showed that agreeable IVAs can more easily build a sense of rapport with a human [45, 46]. In their study, Kang et al. [23] investigated the association between personality traits of human subjects and their feelings when they interacted with an IVA that is incorporated with personality. Their result indicated that agreeable IVAs create stronger rapport especially with agreeable people.

8 Conclusion and Future Work

This study investigated whether the (mis)match in personality between humans and their IVA teammate tends to influence the humans' perception of the IVA's personality. Additionally, this paper studied whether IVA's personality as perceived by humans influenced humans' trust in IVA decisions. While human preference for a particular IVA personality has been previously explored [42, 43], our study went beyond preferences to investigate the influence of (mis)match between the human's and IVA's personalities on the human's perception of the IVAs' personality. Our findings supported the idea that humans are more likely to perceive correctly the personality of the IVA when the personality of their IVA teammate matches their own. Moreover, the humans who perceived the personality of the IVA teammate as agreeable tended to report greater trust in that IVA. As future work, the impact of a (mis)match between humans and IVA teammates on team performance needs to be studied.

References

1. Vosinakis, S., Panayiotopoulos, T.: SimHuman: a platform for real-time virtual agents with planning capabilities. In: de Antonio, A., Aylett, R.S., Ballin, D. (eds.) IVA 2001. LNCS (LNAI), vol. 2190, pp. 210–223. Springer, Heidelberg (2001)
2. Kasap, Z., Magnenat-Thalmann, N.: Intelligent virtual humans with autonomy and personality: state-of-the-art. In: Magnenat-Thalmann, N., Jain, L., Ichalkaranje, N. (eds.) New Advances in Virtual Humans, vol. 140, pp. 43–84. Springer, Berlin (2008)
3. von der Pütten, A.M., Krämer, N.C., Gratch, J.: How our personality shapes our interactions with virtual characters - implications for research and development. In: Allbeck, J., Badler, N., Bickmore, T., Pelachaud, C., Safonova, A. (eds.) IVA 2010. LNCS, vol. 6356, pp. 208–221. Springer, Heidelberg (2010)
4. Doce, T., Dias, J., Prada, R., Paiva, A.: Creating individual agents through personality traits. In: Allbeck, J., Badler, N., Bickmore, T., Pelachaud, C., Safonova, A. (eds.) IVA 2010. LNCS, vol. 6356, pp. 257–264. Springer, Heidelberg (2010)
5. John, O., Srivastava, S.: The Big Five Trait Taxonomy: History, Measurement, and Theoretical Perspectives. Guilford Press, New York (1999)
6. Neto, A.F.B., da Silva, F.S.C.: A computer architecture for intelligent agents with personality and emotions. In: Zacarias, M., de Oliveira, J.V. (eds.) Human-Computer Interaction. SCI, vol. 396, pp. 263–286. Springer, Heidelberg (2012)
7. Bahamón, J.C., Young, R.M.: Toward a computational model for the automatic generation of character personality in interactive narrative. In: Nakano, Y., Neff, M., Paiva, A., Walker, M. (eds.) IVA 2012. LNCS, vol. 7502, pp. 520–522. Springer, Heidelberg (2012)

8. Robison, J., Rowe, J., McQuiggan, S., Lester, J.: Predicting user psychological characteristics from interactions with empathetic virtual agents. In: Ruttkay, Z., Kipp, M., Nijholt, A., Vilhjálmsson, H.H. (eds.) IVA 2009. LNCS, vol. 5773, pp. 330–336. Springer, Heidelberg (2009)

9. Negron, A.P.P., Vera, R.A.A., Jiménez, A.d.A.: Collaborative interaction analysis in virtual environments based on verbal and nonverbal interaction. In: Ninth Mexican International Conference on Artificial Intelligence (MICAI), pp. 129–133 (2010)

10. Prabhala, S.V., Gallimore, J.J.: Designing computer agents with personality to improve human-machine collaboration in complex systems. Wright State University (2007)

11. El-Kassrawy, Y.A.: The impact of trust on virtual team effectiveness. Int. J. Online Mark. 4, 11–28 (2014)

12. Wang, Y., Vassileva, J.: Trust and reputation model in peer-to-peer networks. In: Proceedings of the 3rd International Conference on Peer-to-Peer Computing, p. 150. IEEE Computer Society (2003)

13. Mayer, R.C., Davis, J.H., Schoorman, F.D.: An integrative model of organizational trust. Acad. Manage. Rev. 20, 709–734 (1995)

14. Mitchell, A., Zigurs, I.: Trust in virtual teams: solved or still a mystery? SIGMIS Database 40, 61–83 (2009)

15. Hanna, N., Richards, D.: The impact of communication on a human-agent shared mental model and team performance. In: The 13th International Conference on Autonomous Agents and multi-agent Systems (AAMAS 2014), pp. 1485–1486, Paris, France (2014)

16. Zhao, R., Papangelis, A., Cassell, J.: Towards a dyadic computational model of rapport management for human-virtual agent interaction. In: Bickmore, T., Marsella, S., Sidner, C. (eds.) IVA 2014. LNCS, vol. 8637, pp. 514–527. Springer, Heidelberg (2014)

17. Stanković, I., Popović, B., Focone, F.: Influence of agent behaviour on human-virtual agent body interaction. In: Ronzhin, A., Potapova, R., Delic, V. (eds.) SPECOM 2014. LNCS, vol. 8773, pp. 292–299. Springer, Heidelberg (2014)

18. Bevacqua, E., Stanković, I., Maatallaoui, A., Nédélec, A., De Loor, P.: Effects of coupling in human-virtual agent body interaction. In: Bickmore, T., Marsella, S., Sidner, C. (eds.) IVA 2014. LNCS, vol. 8637, pp. 54–63. Springer, Heidelberg (2014)

19. Rushforth, M., Gandhe, S., Artstein, R., Roque, A., Ali, S., Whitman, N., Traum, D.: Varying personality in spoken dialogue with a virtual human. In: Ruttkay, Z., Kipp, M., Nijholt, A., Vilhjálmsson, H.H. (eds.) IVA 2009. LNCS, vol. 5773, pp. 541–542. Springer, Heidelberg (2009)

20. Neff, M., Wang, Y., Abbott, R., Walker, M.: Evaluating the effect of gesture and language on personality perception in conversational agents. In: Allbeck, J., Badler, N., Bickmore, T., Pelachaud, C., Safonova, A. (eds.) IVA 2010. LNCS, vol. 6356, pp. 222–235. Springer, Heidelberg (2010)

21. Neff, M., Toothman, N., Bowmani, R., Fox Tree, J.E., Walker, M.A.: Don't scratch! Self-adaptors reflect emotional stability. In: Vilhjálmsson, H.H., Kopp, S., Marsella, S., Thórisson, K.R. (eds.) IVA 2011. LNCS, vol. 6895, pp. 398–411. Springer, Heidelberg (2011)

22. Cafaro, A., Vilhjálmsson, H.H., Bickmore, T., Heylen, D., Jóhannsdóttir, K.R., Valgarðsson, G.S.: First impressions: users' judgments of virtual agents' personality and interpersonal attitude in first encounters. In: Nakano, Y., Neff, M., Paiva, A., Walker, M. (eds.) IVA 2012. LNCS, vol. 7502, pp. 67–80. Springer, Heidelberg (2012)

23. Kang, S.-H., Gratch, J., Wang, N., Watt, J.H.: Agreeable people like agreeable virtual humans. In: Prendinger, H., Lester, J.C., Ishizuka, M. (eds.) IVA 2008. LNCS (LNAI), vol. 5208, pp. 253–261. Springer, Heidelberg (2008)

24. Aguilar, R.A., de Antonio, A., Imbert, R.: Searching pancho's soul: an intelligent virtual agent for human teams. In: Electronics, Robotics and Automotive Mechanics Conference (CERMA 2007), pp. 568–571 (2007)

25. Goldberg, L.R.: An alternative description of personality: the big-five factor structure. J. Pers. Soc. Psychol. **59**, 1216–1229 (1990)

26. Van den Bosch, K., Brandenburgh, A., Muller, T.J., Heuvelink, A.: Characters with personality! In: Nakano, Y., Neff, M., Paiva, A., Walker, M. (eds.) IVA 2012. LNCS, vol. 7502, pp. 426–439. Springer, Heidelberg (2012)

27. Barry, B., Stewart, G.L.: Composition, process and performance in self-managed groups: the role of personality. J. Appl. Psychol. **82**, 62–78 (1997)

28. McCrae, R.R., John, O.P.: An introduction to the five-factor model and its applications. J. Pers. **60**, 175–215 (1992)

29. Mann, R.D.: A review of the relationships between personality and performance in small groups. Psychol. Bull. **56**, 241–270 (1959)

30. Scherer, K.R.: Personality Markers in Speech. Cambridge University Press, London (1979)

31. Scherer, K.R.: Personality inference from voice quality: the loud voice of extroversion. Eur. J. Soc. Psychol. **8**, 467–487 (1978)

32. Krishnan, V., Foster, A., Kopper, R., Lok, B.: Virtual human personality masks: a human computation approach to modeling verbal personalities in virtual humans. In: Nakano, Y., Neff, M., Paiva, A., Walker, M. (eds.) IVA 2012. LNCS, vol. 7502, pp. 146–152. Springer, Heidelberg (2012)

33. Argyle, M.: Bodily Communication. Routledge, London (1988)

34. Goldberg, L.R., Johnson, J.A., Eber, H.W., Hogan, R., Ashton, M.C., Cloninger, C.R., Gough, H.G.: The international personality item pool and the future of public-domain personality measures. J. Res. Pers. **40**, 84–96 (2006)

35. Gosling, S.D., Rentfrow, P.J., Swann, W.B.: A very brief measure of the big-five personality domains. J. Res. Pers. **37**, 504–528 (2003)

36. Dillenbourg, P., Traum, D.R., Schneider, D.: Grounding in multi-modal task-oriented collaboration. In: The European Conference on Artificial Intelligence in Education, pp. 401–407 (1996)

37. Arellano, D., Varona, J., Perales, F.J., Bee, N., Janowski, K., André, E.: Influence of head orientation in perception of personality traits in virtual agents. In: The 10th International Conference on Autonomous Agents and Multiagent Systems, vol. 3, pp. 1093–1094. IFAAMAS, Taipei, Taiwan (2011)

38. de Sevin, E., Hyniewska, S.J., Pelachaud, C.: Influence of personality traits on backchannel selection. In: Allbeck, J., Badler, N., Bickmore, T., Pelachaud, C., Safonova, A. (eds.) IVA 2010. LNCS, vol. 6356, pp. 187–193. Springer, Heidelberg (2010)

39. Deng, Z., Bailenson, J.N., Lewis, J.P., Neumann, U.: Perceiving visual emotions with speech. In: Gratch, J., Young, M., Aylett, R.S., Ballin, D., Olivier, P. (eds.) IVA 2006. LNCS (LNAI), vol. 4133, pp. 107–120. Springer, Heidelberg (2006)

40. Burgoon, J.K.: Nonverbal signals. In: Knapp, M.L., Miller, G.R. (eds.) Handbook of Interpersonal Communication, 2nd edn. SAGE Publications, Thousand Oaks (1994)

41. Vinciarelli, A., Salamin, H., Polychroniou, A., Mohammadi, G., Origlia, A.: From nonverbal cues to perception: personality and social attractiveness. In: Esposito, A., Esposito, A.M., Vinciarelli, A., Hoffmann, R., Müller, V.C. (eds.) COST 2102. LNCS, vol. 7403, pp. 60–72. Springer, Heidelberg (2012)

42. Isbister, K., Nass, C.: Consistency of personality in interactive characters: verbal cues, non-verbal cues, and user characteristics. Int. J. Hum Comput Stud. **53**, 251–267 (2000)

43. Nass, C., Moon, Y., Fogg, B.J., Reeves, B., Dryer, D.C.: Can computer personalities be human personalities? Int. J. Hum Comput Stud. **43**, 223–239 (1995)

44. Nass, C., Fogg, B.J., Moon, Y.: Can computers be teammates? Int. J. Hum Comput Stud. **45**, 669–678 (1996)
45. Mooradian, T., Renzl, B., Matzler, K.: Who trusts? Personality trust and knowledge sharing. Manage. Learn. **37**, 523–540 (2006)
46. Hanna, N., Richards, D.: "Building a Bridge": communication, trust and commitment in human-intelligent virtual agent teams. In: The Third International Workshop on Human-Agent Interaction Design and Models (HAIDM 2014) at AAMAS2014, Paris, France (2014)

The Effects of Temperament and Team Formation Mechanism on Collaborative Learning of Knowledge and Skill in Short-Term Projects

Mehdi Farhangian[✉], Martin Purvis, Maryam Purvis,
and Tony Bastin Roy Savarimuthu

Information Science, University of Otago, Dunedin, New Zealand
mehdi.farhangian@otago.ac.nz

Abstract. While collaborative learning has long been believed to hold a great value for organizations and classrooms, Modeling this learning in small, short-term project teams is a challenge. This paper describes the development of an agent-based modeling approach that can assist in understanding the collaborative learning of such project teams. A key aspect of the presented approach is our distinction between knowledge and skills required for the achievement of project goals. Both of these forms of intelligence need to be learned in the project context, but the rate of their expansion or enhancement may proceed differently, depending on the personality makeup of the team and the mechanism employed for team assembly. Based on reports from the theoretical and empirical literature, we derive a multi-agent computational model that characterizes how knowledge and skills may be learned among team members with varying personality attributes. Also, Group formation in virtual learning environments is either done voluntary or with the support from the system. In this connection, we studied two types of group formation mechanisms and the role of each mechanism in the collaborative learning and performance of teams.

Keywords: Knowledge · Skill · Collaborative learning · Multi-agent based simulation · Team formation

1 Introduction

Unlike traditional teams where employees learn and improve their performance through formal training, in many modern projects, collaborative learning within small teams often is undertaken and these teams may be assembled only for specific, short-term tasks. Some examples of these temporary teams include crowdsourcing platforms, scientific collaboration teams, open source software development teams, online games and so on. Also, there has been growing interest in the virtual learning communities where groups of students enhance their learning using Computer-Supported Collaborative Learning (CSCL) environments. How well these teams collaborate and fulfill, their missions will depend on the personalities of the individual team members and how well they can share their knowledge and skills. In this paper, we discuss how team formation mechanisms are involved in the acquisition and retention of skill and knowledge.

© Springer International Publishing Switzerland 2015
F. Koch et al. (Eds.): CARE-MFSC 2015, CCIS 541, pp. 48–65, 2015.
DOI: 10.1007/978-3-319-24804-2_4

In the context of team learning, we believe that there is a significant difference between **knowledge** and **skill**. Knowledge, which can be characterized as "know-what", is articulable, i.e. it can be expressed in linguistic form and transmitted to others relatively easily. On the other hand, a skill, which can be characterized as "know-how", refers to a capability of effective interaction with the environment via a tight feedback loop. Skills, for example, the skill of riding a bicycle, are not easily put into words, since they involve tight feedback loops with the environment; and hence they are not as easily transferred when compared to knowledge. To learn a skill often requires close observation and collaboration with a master who already has the skill.

The goal is to construct a plausible simulation model to provide a prediction of knowledge and skill acquisition and retention in collaborative learning systems where temporary teams are formed for different tasks. This simulation tool could help researchers, managers and teachers to have a better understanding about the effect of group formation mechanisms on collaborative learning. The rest of this paper is organized as follows. In the following section, we review the relevant literature concerning the collaborative learning and team formation. Then, we describe the specifics of our model – both its conceptual elements and its computational aspects. Then, we describe how this model has been implemented algorithmically for agent-based simulation and report on some example results so far obtained.

2 Background

Collaborative learning is a learning method that helps people to retain, transfer, and receive knowledge and skill through intra-group collaboration and competition between groups [1]. The knowledge necessary for performing a task may be declarative, procedural, or a mixture of these two. Declarative knowledge represents factual information; procedural knowledge indicates task knowledge.

Today agent and agent-based services facilitate collaborative learning in crowdsourcing platforms and computer-supported Collaborative Learning (CSCL) environments. Agents can provide decision support for managers or teachers and assist them for some tasks, such as group formation. Designing a real multi-agent tool often entails high cost, time and effort. In this paper we simulate collaborative learning to analyze the effect of attributes such as the team formation mechanism and personality on the performance, knowledge, and skill growth of team members. The existing simulation models and tools such as [2–4] do not cover the personality along with knowledge and skill that are the main focuses of this paper.

ACT-R [5] is a cognitive structure that provides mechanisms representing procedural and declarative knowledge learning and forgetting. We chose to use ACT-R to represent employees or learners memory for acquisition and retention of declarative and procedural knowledge because other similar architectures such as Soar [6] and EPIC [7] are more restricted. Soar does not provide a forgetting mechanism, and EPIC does not provide a rule learning mechanism. A complementary approach to the cognitive approach, such as in the studies above, is to apply agent-based models to simulate human behavior instead of supporting this behavior [8].

Teams may benefit from the way they share information and collaborate, and this aspect of project team performance – how it evolves given the circumstances of personality makeup, skills, and knowledge – has not been explored much extent. In this paper, by employing ACT-R as architecture that deals with the emulation of human mental processes in conjunction with our proposed agent-based model, we describe and simulate our study in this area.

To pursue our examination along this course, one needs to have a reliable characterization of human personality. There are several schemes that have been developed over the years such as Five Factor Model (FFM) [9] however, we believe that the one for which there is the most accumulated data is the Myer-Briggs Type Indicator (MBTI) scheme [10]. This is based on a psychological type scheme originally developed by Carl Jung and modified by Myers [11] and has four personality dimensions: (a) iNtroverted-Extraverted, (b) Sending-iNtuitive, (c) Thinking-Feeling, and (d) Perceptual-Judgemental (the names representing extremal ends of each dimension).

- **Extraversion** vs. **Introversion**– an introverted keeps more to him or herself or faces and an extraverted outer social world.
- **iNtuition** vs. **Sensing**– An intuitive type is more abstract and understands according to his or her inner compass, while a sensor gathers information that is in concrete, objective form.
- **Thinking** vs. **Feeling**– A thinker makes decisions based on logic and demonstrable rationality, and a feeler is more empathetic and attempts to see things from given perspectives a.
- **Judgmental** vs. **Perceptive**– A judger wants things settled and organized, and a perceiver is flexible and spontaneous.

In the following section, we describe our agent-based model that incorporates personality type along with the knowledge and skill levels for each agent. The personality type is assumed to be fixed while the knowledge and skill levels are dynamic.

3   The Model

Figure 1 shows a schematic diagram of an individual agent that works on a project team. It has personality, skill, and knowledge components. Within the knowledge, the component is the "Knowledge Credibility" subcomponent, which stores the confidence in which knowledge sources and interactive partners are held.

The goal is to use this as a modifiable template for the examination of dynamic knowledge and skill influences on individual and team performance via simulation experiments. Agents are seeded with various personality types, knowledge, and skills (as described below), and then simulations are run to examine collaborative learning. For each simulation cycle, agents team up and start working on a task. They exchange what knowledge they have with teammates and update their Knowledge-Credibility values with respect to their teammates. They also improve their skills by observing and imitating their teammates' behaviours.

In the following subsections, further details concerning the operation of these agent components are provided.

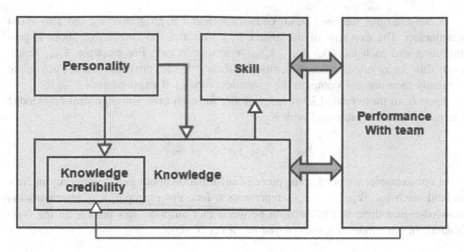

Fig. 1. Model components' overview.

3.1 Task Performance

In our model, each group task needs a set of knowledge and skills. *TASK* is a set of tasks that we have in the system.

$$TASK = \{task_1, task_2, \ldots, task_n\} \tag{1}$$

And each $task_b$ is a vector of l- dimensions; each dimension represents the requirements for that task. And each *task* requires a vector of skill requirements:

$$REQ_b = \{requirement_{b1}, requirement_{b2}, \ldots, requirement_{bn}\} \tag{2}$$

For example, we have a task that is about analyzing health economy data in New Zealand. It requires a set of skill requirements as presented as follows:

$$REQ_1 = \{RProgramming, presentation\} \tag{3}$$

Completing a task requires two sets of knowledge (general knowledge and skill-related knowledge). Before the acquisition of one skill, one needs to learn a knowledge set related to that skill: Here K_{rvb} represent the knowledge matrix related to skills for task b.

$$K_{rvb} = \begin{bmatrix} k_{rb11}k_{rb12}\ldots k_{rb1n} \\ k_{rb21}k_{rb22}\ldots k_{rb2n} \\ \cdot \\ \cdot \\ \cdot \\ k_{rbm1}k_{rbm2}\ldots k_{rbmn} \end{bmatrix} \tag{4}$$

In our example, we need some knowledge about R programming and also about presentation. The first row of the matrix K_{rvb} indicates the knowledge about R programming and each $k_{rb11}, k_{rb12}, \ldots, k_{rb1n}$ represents a fact. For example, k_{rb11} represents this knowledge: microbenchmark library in R provides infrastructure to accurately measure and compare the execution time of R expressions.

Apart from these related-knowledge skills, for each task, some general knowledge is required that is represented with K_{gb}.

$$K_{gb} = \begin{bmatrix} k_{gb1}, k_{gb2}, \ldots, k_{gbm} \end{bmatrix} \tag{5}$$

In our example, we need some piece of information about health economy in New Zealand, each $k_{gb1}, k_{gb2}, \ldots, kb_{gbm}$ represents a fact. For example, k_{gb1} represents the knowledge that there is a correlation between diet nutrition and income in the New Zealand. In our model, each employee has a set of skills,

$$skill_i = \{ skill_{i1}, skill_{i2}, \ldots, skill_{in} \} \tag{6}$$

Each element in the $skill_i$ vector represents the qualification of employee. For example, for employee 1, $skill_1$ represents his vector skill that each one represents a specific skill. And $skill_{11}$ represents R programming is 0 and $skill_{12}$ represents that represents MATLAB programming is 5. The competency of members in skills is calculated as follows:

$$Sk_{il} = 1 - \min\{0, |skill_{il} - requirement_{bl}|\} / skill_{il} \tag{7}$$

Sk_{il} indicates the competency of employee i in domain l; $skill_{il}$ indicates the level of skill of employee i in domain l; and $requirement_{bl}$ indicates the level of skill requirement in domain l in task b We used this formula to avoid giving credit to the employees' over qualifications. The sum of the competency of employee i in task b is calculated by the sum of his competency in all the domains as follows:

$$Sk_{ib} = \sum_{l=1}^{m} Sk_{il} \tag{8}$$

Sk_{ib} represents competency of employee i in task b, and m represents the number of domains in the task b for employee i.

Also, each employee has some knowledge vectors for each skill that is represented as following matrix:

$$K_{evi} = \begin{bmatrix} k_{ei11} k_{ei12} \ldots k_{ei1n} \\ k_{ei21} k_{ei22} \ldots k_{ei2n} \\ \cdot \\ \cdot \\ \cdot \\ k_{eim1} k_{eim2} \ldots k_{eimn} \end{bmatrix} \tag{9}$$

K_{evi} represent the knowledge vector related to each skill for employee i.

Apart from knowledge related to skill, each employee has two other knowledge vector including general knowledge and knowledge about other people.

$$K_{gi} = [k_{gi1}, k_{gi2}, \ldots, k_{gim}] \qquad (10)$$

K_{gi} represents the general vector of employee i. And K_{ij} in the following vector represents the knowledge of employee i about knowledge credibility of employee j.

$$K_{ij} = [k_{i1}, k_{i2}, \ldots, k_{ij}] \qquad (11)$$

The final performance of the employees in the tasks is related to their skill competency and general knowledge competency. In our example skills in R programming and presentation and also knowledge about health economy improve the task performance. Knowledge competency is calculated as follows:

$$KK_{gib} = \max\{0, K_{gb} - K_{gi}\} \qquad (12)$$

As having both of these factors, is critical for the performance a task, following the formula is suggested:

$$Pe_b = \left(\sum_{i=1}^{n} W_{si} * Sk_{ib}\right) * \left(\sum_{i=1}^{n} W_{ki} * KK_{gib}\right) \qquad (13)$$

Pe_b indicates the performance of a team in task b, Sk_{ib} indicates the competency of agent i for task b, and Kk_{gib} indicates the general knowledge competency of agent i for task b. Also, W_{si} indicates the importance of skill i and W_{ki} indicate the importance of knowledge i.

In the rest of the paper, we argue that skill and knowledge improve over time and demonstrate how personalities of employees make a difference in employees' leaning and teams' performances.

3.2 The Influence of Personality

In our model, there are three personality dimensions (as specified by the MBTI scheme) that come into play. Associated with these three personality dimensions, six assumptions are considered and as explained as follows. These assumptions are based on studies reported in the literature about MBTI and team behavior [12–16].

- 1st assumption: Compared to Feeling types, a Thinker's relationship with a person is more sensitive to their knowledge of that person.
- 2nd assumption: Sensors record the result of their satisfying or unsatisfying team experiences as facts more than iNtuitive types do.
- 3rd assumption: Sensors have a higher rate of gathering knowledge from others compared to iNtuitive types.
- 4th assumption: iNtuitive types have a higher rate of self-learning knowledge compared to the Sensors.

- 5th assumption: It is more likely for extraverted types to share their knowledge compared to introverted types.
- 6th assumption: Introverted types have a higher self-learning rate compared to Extraverted types.

A number represents the degree of personality in each dimension is presented as follows:

- Introverted/Extraverted (IE): (range 0–$0.5 \rightarrow$ *Introverted* $0.5 - 1 \rightarrow$ *Extraverted*)
- iNtuitive/Sensing(NS):(range 0–$0.5 \rightarrow$ *Intuitive* $0.5 - 1 \rightarrow$ *Sensor*),
- Thinking/Feeling (TF):(range 0–$0.5 \rightarrow$ *Feeler* $0.5 - 1 \rightarrow$ *Thinker*),
- Perceiving/Judging (PJ): (range 0–$0.5 \rightarrow$ *Perceiver* $0.5 - 1 \rightarrow$ *Judgers*).

Apart from personality variables, some other non-personality variables affect decisions and behaviour. These factors are discussed in the following sections. These factors include task performance, knowledge credibility, knowledge growth, skill growth and forgetting (of both knowledge and skill).

3.3 Knowledge Sharing

Knowledge can be shared through communication. In our knowledge-sharing model, two main factors including having a common goal (being in one group) and desire to have connections with others (extraversion) can cause more knowledge sharing.

As mentioned in the *5th assumption*, extraverted types are more likely to share their knowledge compared to introverted types, who limit their social activities to a few people. So the probability of sharing knowledge with another agent is related to two factors. IE_i (Level of Extraversion of agent) and In_i (in-group factor that is a binary value if agent j is in same group, $In_j = 1$, or if an agent is in another group, $In_j = 0$). The probability of sharing knowledge calculated as follows:

$$Sh_{ij} = \frac{w_{IE}IE_i + w_{In}In_j}{w_{IE} + w_{In}} \tag{14}$$

Where Sh_{ij} is agent i probability of sharing knowledge with agent j. And weights w_{IE} indicates w_{In}, indicate the importance of Extraverted personality, In-group factor respectively. The willingness to accept shared knowledge is related to Knowledge-credibility (trust), and it is explained in the next section.

3.4 Trust (Knowledge Credibility)

Trust is **a crucial** part of knowledge sharing [17]. The knowledge-sharing process entails two different socio-cognitive decisions [18]:

(1) a decision to pass or not pass on a piece of knowledge
(2) a decision to accept or reject a given piece of knowledge.

The degree of confidence that one has in the integrity and competence of the organizational environment is essential for both of these decisions [17].

Although trust can take different forms, we assume in our organizational context here that trust refers to the degree to which a person can have confidence in the information that he or she may receive from a coworker; and we call it knowledge-credibility. There are three principal routes by which we can acquire information relevant to team performance: team success, direct interaction, and indirect interaction:

(1) **Team success:** This parameter reflects the history of previous team successes.
(2) **Direct Interaction:** agents gather information from the expertise of another agent who shares his knowledge.
(3) **Indirect interaction:** each agent gathers third parties' attitudes about other agents. The average of these attitudes determines the general reputation of the agent.

As a result overall the Knowledge-credibility of agent i on agent j is calculated as follows:

$$Kc_{ij}(t) = \frac{(w_{Id}Id_{ij}(t) + w_{Re}Re_{ij}(t) + w_{Ts}Ts_{ij}(t))}{w_{Id} + w_{Re} + w_{Ts}} \tag{15}$$

Kc_{ij} refers to Knowledge-credibility of agent i to agent j at time t This knowledge-credibility is affected by three factors: Ts_{ij} (team success), Id_{ij} (direct interaction), and Re_{ij} (indirect interaction or reputation). Weights w_{Id}, w_{Re}, w_{Ts} determine the importance of direct trust, indirect trust and team success, respectively. These three factors are explained in the following sections.

3.4.1 Team Success

Team success reflects agents' past team experiences with other agents and represents the total number of satisfying and successful group tasks.

If the performance of the task is less than the threshold, θ_1 the task is unsatisfying. Otherwise it is satisfying. Agents update their belief about team members after each task by this formula:

$$Ts_{ij}(t) = \begin{cases} Ts_{ij}(t-1) + e^{NS_i w_{NS}} Pe_{ijb}/100 & \text{if } Pe_{ijb} > \theta_1 \\ Ts_{ij}(t-1) - \frac{e^{NS_i w_{NS}}}{100 Pe_{ij}} & \text{otherwise} \end{cases} \tag{16}$$

$Ts_{ij}(t)$ indicates the belief of agent i about past experience with agent j. NS_i represents the sensing personality of agent i, and Pe_{ijb} represents the performance in task b where agent i and agent j are team members. As mentioned above in the *2nd assumption*, for people with a Sensing personality, what happened in the past is a more important factor compared to iNtuition types, and w_{NS} indicates the importance of the Sensing personality on team success factor on Knowledge-credibility.

3.4.2   Direct Interaction

Over the course of time, agents update their beliefs about other agents' expertise and develop their Knowledge-credibility. If agent j shares some knowledge with agent i, agent i develops his belief on (confidence in) the expertise of agent j as described in the following formula:

$$Id_{ji}(t) = \begin{cases} Id_{ji}(t-1) - w_{TF}(1-TF_i) & K_j = 0 \; and \; K_i = 1 \\ Id_{ji}(t-1) + w_{TF}(1-TF_i) & if \; Agent \; i \; accept \; K_j \\ Id_{ji}(t-1) & otherwise \end{cases} \qquad (17)$$

$Id_{ji}(t)$ indicates the direct trust of agent j on agent i; TF_i indicates the degree of feeling personality of agent i; and $(1-TF_i)$ determines thinking of this agent. And w_{TF} indicates the weight of thinking-feeling dimension. In this formula we face 3 scenarios which are based on the 1st Assumption (above):

(1) If agent j expresses his opinion about a topic on which he does not have any knowledge (i.e. $K_j = 0$, then it would have a negative effect on agent $i's$ opinion who knows that's j is wrong. Agent i decrease his value of Knowledge-credibility based on his thinking-feeling personality. People with thinking personality make judgements based on empirical verification, so it makes them more sensitive to false knowledge.
(2) Agent i may accept the knowledge from agent j. The details about accepting knowledge are explained in the knowledge sharing section.
(3) Agent i may receive knowledge from agent j and without knowing whether the knowledge is true or false. In this case it will not have any effect on agent $j's$ Knowledge-credibility.

3.4.3   Indirect Trust (Reputation)

Agents not only compute Knowledge-credibility based on expertise and team success, but also, they collect recommendations from other agents. When agent l interacts with agent i and transfers his attitude towards a third party, agent j, he is building agent j's reputation for agent i. So the reputation of agent j is calculated as follows:

$$Re_{ij}(t) = Re_{ij}(t-1) + Kc_{il}(t) * Kc_{lj}(t) \qquad (18)$$

$Re_{ij}(t)$ indicates the reputation of agent j for agent i at time t. $Kc_{il}(t)$ indicates the knowledge credibility of agent i to agent l, and $Kc_{lj}(t)$ indicates the knowledge credibility of agent l to agent j. The way, which people exchange information about other agents is similar to knowledge sharing that is explained in Sect. 3.4.

3.4.4   Knowledge Acceptance

As mentioned earlier the willingness to accept shared knowledge is related to Knowledge-credibility (trust). This is relevant to sensing personality as mentioned in the 3rd assumption. When agent i share his knowledge with agent j, the probability that agent j accepts the knowledge is related to his Knowledge-credibility and Sensing personality. In the MBTI scheme, people with Sensing personalities are more willing to gather facts compared to iNtuition types.

The probability that knowledge is accepted by agent j is calculated as follows:

$$a_{ji} = \left(w_{Kc}e^{Kc_{ji}}/10 + w_{NS2}NS_j\right)/(w_{KC} + w_{NS2}) \qquad (19)$$

a_{ji} is agent j willingness to accept knowledge from agent i that is related to two factors: Kc_{ji} (the Knowledge-credibility of agent j for agent i) and NS_j (the level of Sensing in agent j). Where weights w_{Kc}, w_{NS2} indicate the importance of Knowledge-credibility and the Sensing personality in accepting knowledge, respectively.

3.5   Self-learning Knowledge

In addition to learning skill from others, we cover the effect of self-learning. In each time step, people increase their knowledge at a rate that is related to the Introverted and iNtuition components of their personalities. Introverted types have a higher self-learning rate than Extraverted types (6th assumption), and iNtuitive types can generate new knowledge by interpreting their past knowledge (4th assumption).

This probability is calculated as follows:

$$Sl_i = \frac{\theta_5(w_{IE2}(1 - IE_i) + w_{NS3}(1 - NS_i))}{w_{IE2} + w_{NS3}} \qquad (20)$$

where Sl_i indicates the probability of self-learning of agent i. Again, this probability determines the likelihood of a knowledge topic's value getting set to a value of 1. IE_i reflects where the agent lies along the Introverted-Extraverted personality dimension, and NS_{i_i} indicates where along the Sensing-iNtuition dimension (values are from 0 to 1). w_{IE2}, w_{NS3} indicate the importance of Introverted and iNtuition personality types, respectively, and θ_5 shows the rate of self-learning knowledge growth.

3.6   Skill Learning

Employees not only learn the knowledge by interacting with other agents; they can also improve their skills or procedural knowledge by observing others' behavior. Observational learning is an effective method of collaborative learning that is commonly used by both human and computer models [19]. In observational learning, people need a model to imitate the behavior. In our model, agents improve their skills by observing and imitating another agent who is using the same skill in their team. Two factors affect the improvement of skill – the difference between the skills of people who are performing the task and the amount of relevant knowledge that the learner has. In our simulation model, skill improvement of an agent is calculated as follows:

$$skill_{iv}(t) = skill_{iv}(t - 1) + K_{evi}\theta_2(skill_{iv}(t - 1) - skill_{-iv}(t - 1)) \qquad (21)$$

Skill improvement is affected by K_{evi} which represents the sum of knowledge related to $skill_{iv}$. And $skill_{iv}(t)$ indicates the skill v of agent i in time t, and θ_2 shows the growth rate of skill. $skill_{-iv}$ indicates the skill v of other members in the team.

3.7 Forgetting

People forget their knowledge and skills if they stop using them, but the degree of forgetting differs in knowledge and skill. In order to model how people learn and forget knowledge and skill, we used declarative and procedural memory that is presented in the ACT-R cognitive architecture [20]. In this model, declarative knowledge represents factual information, and procedural knowledge indicates task knowledge.

In ACT-R, a declarative memory item is dependent on how often (frequency) and how recently (recency) the item is used. Also in the higher stages of learning, the strength of declarative memory increases by practicing. However, when knowledge is stored in procedural memory, it will not easily decay with time.

In our model, we assume that knowledge is stored in the declarative memory and skill is stored in the procedural memory. The forgetting rate in knowledge is faster than skill but also depends on the competency of agents in that skill. So, skill deterioration (when employees are not using that skill) is calculated as follows:

$$skill_{iv}(t) = skill_{iv}(t-1) - \theta_3 e^{-(skill_{iv}(t))} skill_{iv}(t-1) \tag{22}$$

$skill_{iv}(t)$ indicates the skill v of agent i in time t, and θ_3 shows the forgetting rate of the skill.

In addition to frequency and recency, which are mentioned for skill forgetting, the competency in the skill related to that knowledge reduces the forgetting rate of knowledge [21].

Each time that a person uses knowledge; this knowledge is refreshed and is saved from forgetting. The probability that a person loses his knowledge is related to the strength of skill related to this knowledge. So, the probability of forgetting knowledge is calculated as follows:

$$P_{fk} = \theta_4 e^{-(skill_{iv}(t)))} \tag{23}$$

P_{fk} indicates the probability of forgetting knowledge, $skill_{iv}(t)$ indicates the competency in the skill related to knowledge, and θ_4 indicates the rate of knowledge forgetting.

4 Simulation

The proposed mathematical model was translated into an agent-based model and implemented in Repast [22]. In this model, self-organizing teams perform a task in the context of a temporary project. Each temporary project consists of two tasks, and each task is related to a single skill, and two people are required to work on a task. So, a temporary project needs four employees.

Initial setup of the experiment comprised 100 employees and 25 tasks, with each task requiring four employees. Each individual has some initial properties, such as a vector of skills, a matrix of knowledge related to these skills, and a knowledge credibility vector of other employees. In each cycle, individuals team up and start a task. Each task takes 100 time-steps. In each time, step agents develop their trust of each other and knowledge that is explained in detail in Sect. 4.1 by communicating and updating their skills by observation. In this paper, two task allocation mechanisms are studied: based on trust (knowledge credibility) and skill.

(1) **Knowledge credibility:** In the first scenario, employees form a team based on their knowledge credibility. We assume one employee starts a task and asks three other members with the highest knowledge credibility to join that task.

Initialization: Group formation mechanism and inputs such as task and team members' characteristics

  For each task

    FOR time step < current-time step + 100 **DO**

      FOR each time step

      FOR each agent i

        With probability Sh_{ij} *(formula 12)*:

          Share knowledge with   agent j

        IF receive a knowledge from agent j

          With probability of a_{ij} *(formula 14)*:

            Accept the knowledge

            Update direct interaction $Id_{ji}(t)$ *(formula 8)*

        END IF

        IF receive information about agent k

            Update $Re_{ik}(t)$ *(formula 10)*

        END IF

        With probability Sl_i *(formula 15)*:

          Update knowledge

        When agent i using same skill as agent j

        AND they are in same team:

          Improve skill ΔSk_{ij} *(formula 17)*

        WHEN agent i stops using skill v: forget *(formula 18)*

        WHEN agent i stops using knowledge k : forget

        (formula 19)

      END FOR

      END FOR

    END WHILE

    Calculate team performance Pe_b *(formula 3)*

    IF agent j is in the same team

      Update team success trust $Ts_{ij}(t)$ *(formula 9)*

  Next task allocation

Fig. 2. Pseudo code of the agent simulation model.

(2) **Skill competency:** In the second scenario, people are assigned to a task based on their competency. Managers assign a combination of employees with the highest skill as explained in Formula 7.

Initially for each of the four MBTI personality dimensions, we established a scale between 0 and 1 and assigned values for each employee. In our initial settings, a vector contains 10 knowledge items assigned to each skill. In addition to that knowledge, we have a general vector of knowledge that contains 100 elements. We assume each project needs a maximum of 50 units of this knowledge.

The values assigned 1 for the weight parameter and number 100, 0.1, 1, 10, 1 to the parameters θ_1, θ_2, θ_3, θ_4, and θ_5 respectively and we receive the results of 100 model runs for the model analysis. We ran two types of experiments: firstly, we compared two task allocation mechanisms and their differences in knowledge learning, skill learning, and team performance by assigning a random personality to the agents. Then, we compared the effects of different types of employees (in terms of personality) and their roles in the team performances in two task allocation mechanisms (Fig. 2).

Also, we are developing a proposed simulation tool to help managers and teachers identify how changes in knowledge, skill, and the performance of group m embers appear due to their attributes such as personality, skill, knowledge, task requirements, and also the task allocation mechanism. A schematic representation of this tool is illustrated in Fig. 3.

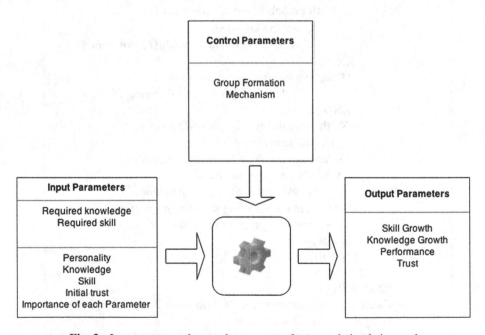

Fig. 3. Input-output and control parameter of proposed simulation tool

5 Results

In our computer simulation, we compared knowledge growth, skill growth, and performance while performing 10 tasks (1000 time steps) using two task allocation mechanism. Figure 4 compares the average knowledge of employees (an average over 100 runs) for both team-formation mechanisms (based on knowledge credibility and skill). Figure 5 shows a comparison of the average skills of employees (averages over 100 model runs) for both team-formation mechanisms – based on credibility and skill-based team formation after 10 tasks (1000 time steps). Figure 6 compares the average team performances (averaged over 100 model runs) for both team formation mechanisms based on credibility and skill-based team-formation after 10 tasks.

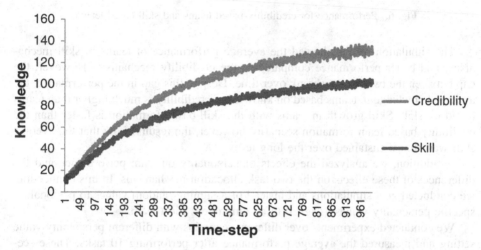

Fig. 4. Knowledge growth for credibility-based teams and skill-based teams.

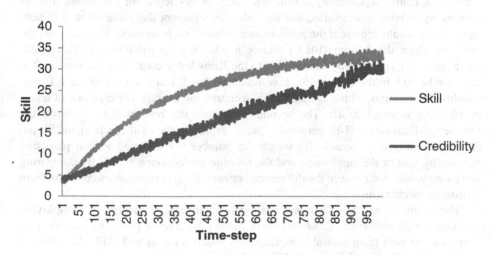

Fig. 5. Skill growth for credibility-based teams and skill-based.

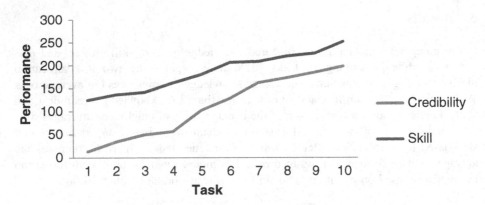

Fig. 6. Performance for credibility-based teams and skill-based teams.

The simulation results showed the average performance of teams in skill mechanisms had better performance compared to the credibility mechanism. However, the gap between the two results shrank over time. Despite this gap in the performance, the average knowledge in teams based on knowledge credibility is much higher than teams based on skill. Skill growth in teams with the skill-based formation is faster than the credibility-based team formation scenario; however, the results show that the average skill was almost sustained over the long term.

In addition, we analyzed the effects of personality on team performance and the differences of these effects on the two task allocation mechanisms. In this connection, we conducted new simulations and instead of assigning random values to personality, specific personality values assigned to all employees for a team.

We conducted experiments over different scenarios with different personality value setting and measured the average performance after performing 10 tasks. These scenarios were measured for two self- rized in Fig. 5, which shows a heat map, with each value of a matrix representing a different color. Rows represent the dimensions of personality in both mechanisms, and the columns represent the value of each dimension.. These results represent the performance value of each scenario. For example, the first row from the bottom (I-E-C) shows a particular distribution of Introverted-Extraverted (I-E) personality with respect to the Knowledge-credibility mechanism (C). The number 0.1 in the Personality axis indicates that 0.1 is assigned to the I_E personality dimension of all the agents. In this scenario, the average performance of teams in 10 tasks is equal to 10. The second row from the bottom (I-E-S) shows the Introverted-Extraverted (I-E) personality with respect to the skill mechanism (S) and the first number is a scenario for which the number 0.1 assigned to that particular personality trait of the employees, and the average performance was 8. By comparing · these two values, we observe the difference between team performances based on team formation mechanisms.

The results reveal that, there is a relationship between personalities of employees and the overall performance. Results show Extraverts have a positive effect on performance for both team assembly mechanisms based on trust and skills. However, a balance of Introverts and Extraverts led to a better result compared to the scenarios for which all members are very Extraverted. The observed behavior showed increasing Extraversion had a positive effect in the Skill-based scenarios compared to the Knowledge-credibility-based scenarios. In the other words, if team members are skillful, some teams' member with a particular (such as being Extraverted) could end up with more knowledge-sharing and consequently improved performance.

Sensing-iNtuition personalities have almost opposite effects on the two team-formation mechanisms, and they follow different patterns. Intuition is a more important factor in Knowledge credibility-based teams compared to skill-based teams. A simple, approximate explanation of this behavior is as follows. First, in a system where all the employees are Sensors, they are eager to gather additional knowledge. Since teams are formed based on credibility, this virtue assists them for a high knowledge sharing rate. When team formation is based on skill and employees are intuitive, they do not share their knowledge and this phenomenon results in negative learning and consequently poor performance.

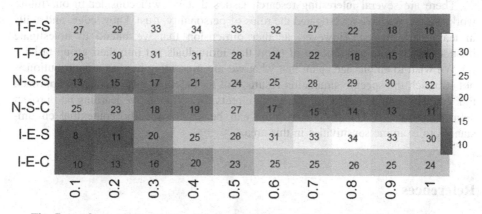

Fig. 7. performance and personality in credibility-based teams and skill-based teams.

Having a high Thinking personality was shown to be better in our simulations than having a high Feeling type of personality in most of the cases. The Thinking personality had more advantages for team formation based on knowledge credibility compared to team formation based on skill. This reflected the effect that when people have thinking personalities and team formation is based on knowledge credibility; they eventually find better teams to work with. When people are feelers they might trust in wrong persons and give them the credit that they do not deserve that but in a world with thinking people these mistakes less likely occur.

6 Discussion and Future Work

The growth of collaborative learning in crowdsourcing platforms and CSCL systems suggests that a simulation environment could provide better understanding of group formation and learning process. In this paper, we have developed a model that shows how people in collaborative learning systems may grow their knowledge and skill via collaborative learning. Group formation in virtual learning environments is either done voluntary or with the support from the system. We investigated how a group formation mechanism might affect the collaborative learning and team performance. So, we compared the results of two group formation mechanisms: based on skill and based on knowledge credibility.

The results of our simulations showed that although team assembly based on skill ended up with good performance, they are not necessarily successful in collaborative learning. In particular, knowledge increased more in the credibility-based team-formation mechanism. We also investigated the effects of temperament (personality) on team performance for both team-assembly mechanisms, and we observed several interesting results as summarized in Fig. 7.

Implication derived from the simulation environment could provide a low cost tool for managers, teachers and researchers to have a better understanding of the impacts of different scenarios on teams' collaborative learning and performance.

There are several interesting research issues that we will consider in our future work. So far, we have investigated the roles of personality, trust, knowledge, and skills in team performance. However, another dimension that we intend to investigate includes motivation and amount of effort that individuals put into their group tasks.

We wish to emphasize again that what we are presenting here as a contribution is not so much the specific simulation results, but a modelling and simulation approach that can demonstrate interesting emergent effects for collaborative learning and project team performance. The parameterization can be set for specific contextual circumstances to examine sensitivities in this area.

References

1. Chen, L., Yang, Q.: A group division method based on collaborative learning elements. In: The 26th Chinese Control and Decision Conference (2014 CCDC), pp. 1701–1705 (2014)
2. Spoelstra, M., Sklar, E.: Using simulation to model and understand group learning. In: Proceedings of AAMAS, vol. 7 (2007)
3. Khandaker, N., Soh, L.-K.: ClassroomWiki: a wiki for the classroom with multiagent tracking, modeling, and group formation. In: Proceedings of the 9th International Conference on Autonomous Agents and Multiagent Systems: volume 1-Volume 1. International Foundation for Autonomous Agents and Multiagent Systems (2010)
4. Khandaker, N., Soh, L.-K.: SimCoL: a simulation tool for computer-supported collaborative learning. IEEE Trans. Syst. Man Cybern. Part C Appl. Rev. 41(4), 533–543 (2011)
5. Anderson, J.R., Bothell, D., Byrne, M.D., Douglass, S., Lebiere, C., Qin, Y.: An integrated theory of the mind. Psychol. Rev. 111(4), 1036–1060 (2004)
6. Laird, J.: SOAR: an architecture for general intelligence. Artif. Intell. 33(1), 1–64 (1987)
7. Kieras, D.E., Meyer, D.E.: An overview of the EPIC architecture for cognition and performance with application to human-computer interaction. Hum. Comput. Interact. 12(4), 391–438 (1997)

8. Martínez-Miranda, J., Pavón, J.: Modelling trust into an agent-based simulation tool to support the formation and configuration of work teams. In: Demazeau, Y., Pavón, J., Corchado, J.M., Bajo, J. (eds.) 7th International Conference on Practical Applications of Agents and Multi-Agent Systems (PAAMS 2009). AISC, vol. 55, pp. 80–89. Springer, Heidelberg (2009)
9. Costa, R.R.M., Paul T.: Professional manual: revised NEO personality inventory (NEO-PI-R) and NEO five-factor inventory (NEO-FFI). In: Psychological Assessment Resources, Odessa, FL (1992)
10. Myers, I.: The Myers-Briggs Type Indicator. Consulting Psychologists Press, Palo Alto (1962)
11. Jung, C.G.: Psychological Types: Or the Psychology of Individuation. Harcourt, Brace, New York (1921)
12. Varvel, T., Adams, S.G., Pridie, S.J., Ruiz Ulloa, B.C.: Team effectiveness and individual Myers-Briggs personality dimensions. J. Manage. Eng. 20(4), 141–146 (2004)
13. Myers, I.B., Most, R., McCaulley, M.H.: Manual: A Guide to the Development and Use of the Myers-Briggs Type Indicator. Consulting Psychologists Press, Palo Alto (1985)
14. Capretz, L.F.: Personality types in software engineering. Int. J. Hum Comput Stud. 58(2), 207–214 (2003)
15. Bradley, J.H., Hebert, F.J.: The effect of personality type on team performance. J. Manage. Dev. 16(5), 337–353 (1997)
16. Cruz, S., da Silva, F.Q.B., Capretz, L.F.: Forty years of research on personality in software engineering: a mapping study. Comput. Human Behav. 46, 94–113 (2015)
17. Dignum, V., Van Eijk, R.: Towards a model to understand the influence of trust in knowledge sharing decisions. In: Workshop on Trust AAMAS (2005)
18. Castelfranchi, C.: Trust mediation in knowledge management and sharing. In: Jensen, C., Poslad, S., Dimitrakos, T. (eds.) iTrust 2004. LNCS, vol. 2995, pp. 304–318. Springer, Heidelberg (2004)
19. Fernlund, H.K.G.: Evolving models from observed human performance. Dissertation, University of Central Florida Orlando, Florida (2004)
20. Anderson, J.R.: Learning and Memory: An Integrated Approach. Wiley, New York (2000)
21. Kim, J.W., Ritter, F.E., Koubek, R.J.: An integrated theory for improved skill acquisition and retention in the three stages of learning. Theor. Issues Ergon. Sci. 14(1), 22–37 (2013)
22. University of Chicago and Argonne National Laboratory: RECURSIVE Porous Agent Simulation Toolbox (2007)

Exploring Smart Environments Through Human Computation for Enhancing Blind Navigation

Hugo Paredes[1(✉)], Hugo Fernandes[1], André Sousa[1], Luis Fernandes[1], Fernando Koch[2], Renata Fortes[3], Vitor Filipe[1], and João Barroso[1]

[1] INESC TEC and Universidade de Trás-os-Montes e Alto Douro,
Vila Real, Portugal
{hparedes,hugof,andresousa,lfernandes,
vfilipe,jbarroso}@utad.pt
[2] SAMSUNG Research Institute, Campinas, SP, Brazil
fernando.koch@samsung.com
[3] Computer Science Department (SCC),
University of Sao Paulo (USP), São Carlos, SP, Brazil
renata@icmc.usp.br

Abstract. In this paper the orchestration of wearable sensors with human computation is explored to provide map metadata for blind navigation. Technological navigation aids for blind must provide accurate information about the environment and select the best path to reach a chosen destination. Urban barriers represent dangers for the blind users. The dynamism of smart cities promotes a constant change of these dangers and therefore a potentially "dangerous territory" for these users. Previous work demonstrated that redundant solutions in smart environments complemented by human computation could provide a reliable and trustful data source for a new generation of blind navigation systems. We propose and discuss a modular architecture, which interacts with environmental sensors to gather information and process the acquired data with advanced algorithms empowered by human computation. The gathered metadata should enable the creation of "*happy maps*" that are delivered to blind users through a previously developed navigation system.

Keywords: Blind navigation · Crowdsourcing · Human computation

1 Introduction

Visual impairment poses some restrictions and special requirements to human mobility and urban barriers, in particular, represent dangers for the blind. The dynamism of smart cities is prone to create constant changes of these dangers and is therefore a potentially "dangerous territory" for these users [1]. Previous work [10] demonstrated that solutions with multiple sensing input sources in smart environments, complemented by human computation, could provide a reliable and trustful data source for a new generation of blind navigation systems. However, most of these solutions require constant user feedback causing meshing and hindering the use of the devices. Some studies have also demonstrated the need to include pervasive technology to support

© Springer International Publishing Switzerland 2015
F. Koch et al. (Eds.): CARE-MFSC 2015, CCIS 541, pp. 66–76, 2015.
DOI: 10.1007/978-3-319-24804-2_5

navigation and guidance. These same technologies may also be used for the acquisition of environmental and user data.

Typical technological navigation aids for the blind are designed to provide accurate information about the environment and select the best path to reach a chosen destination. However, with the constant development of sensing technologies and ubiquitous computing new types of information sources can be used to create a new type of navigation systems for the blind. Some approaches have considered the use of crowdsourcing to give inputs about unexpected dangers in the route [2–5]. However, the role of extended geographical information data, like route appealing or safeness (Happy Maps [1]), has not been considered when recommending routes. Sometimes the shortest path is not the ideal path to get to a destination.

This paper proposes and discusses a conceptual architecture, which interacts with environmental sensors to gather information and process the acquired data with advanced algorithms empowered by human computation. The gathered metadata should enable the creation of "happy maps" that are delivered to blind users through a previously developed navigation system. As it is intended to be a non-intrusive system, user components are embedded in objects/devices that are part of the daily lives of users, in particular, the white cane, the watch, the glasses or the phone. In most cases consumer electronics devices, mass produced, can be used and thus are more economic. These are the cases of smartwatches, smartphones and smartglasses, which can be adapted for use by the blind, customized for their needs and capabilities. In specific situations, as for the white cane, hardware customization is required, following previous work [6]. With these devices it is possible to obtain several data from the sensors, which once contextualized, can allow the inference of environmental information. To ensure the dynamism of the solutions it appears to be necessary to apply artificial intelligence techniques to get the association with known objects/obstacles. However, such mechanisms represent no solution to the time variation of the environment, whereby the model specifies retro-feedback mechanisms for dynamic updating of the information, and also acting as a validation mechanism. The presented conceptual architecture is faced with the future challenges of the navigation systems for the blind, setting some of the future guidelines for these systems.

Section 2 presents some background on the topics involved. In Sect. 3 the model for enhancing blind navigation based on the use of crowdsourcing and ubiquitous sensing and computing to generate happy maps is proposed. An application scenario is presented in Sect. 4 and the discussion of the proposal follows up in Sect. 5. Finally, Sect. 6 concludes with some final remarks.

2   Background

In a recent survey, Zhu et al. [8] argue that mature infrastructures for sensing data generation, collection, classification, analysis, and processing are desired. Cloud computing is essential to build high performance platforms. Currently, sensor networks are usually restricted to small regions. However, in the near future they should be at the town or city level, or even world level. Data will be aggregated and distributed in different methods to all potential users. Internet of Things (IoT) is the forerunner of building such large-scale networks and it is one of the top concerned research topics. Data management methods will bloom as well as other data sciences to solve problems in the world of big data, such as data management, security and privacy, data protection and integrity, machine learning, neural networks, data mining and crowd sourcing.

2.1 Smart Cities and Accessibility

In the last years some solutions of crowd participation in Smart Cities and Accessibility have been explored [2, 3]. In order to improve a city living experience, it is necessary to know its problems, and no one knows those problems better than the citizens. Applications that empower citizens to report the city problems can contribute to enhance the quality of life by prioritizing the gathered information and solving the most relevant situations. However this evaluation and prioritization should take into consideration the way people travel, how those obstacles force them to make changes in their path and how the population is affected by those changes. Therefore, if these users' emotions can be interpreted, then those obstacles can be classified according how they affect their mood. The use of specific applications can also contribute to this evaluation by analyzing and inferring on the user context to retrieve specific information without the need to ask, for example the name of the streets. Information like the sidewalk length, preferred paths, (not just the fastest, but also the most likable or safest one), fix or new obstacles, (e.g. new construction in progress) should be considered when creating a happy map. These solutions are based on collecting smartphone data (via a specific application like the IBM Accessible Way[1], and others), store the data on a server and use models to identify not only the reported problems, but also their impact [4]. This information can be used to help people in their everyday life, especially in the case of blind users. They can be warned that the sidewalk is inappropriate for walking safely, or that the sidewalk has many holes or obstacles blocking the way. If a representative amount of data is collected through people engagement, a significant number of issues will be registered, and analytics tools can be used to identify the impact of these issues on the life of people with disabilities. Consequently, this information can be used to build accessibility maps, and to define an appropriate action plan to address the detected issues, optimizing resource allocation [5]. Open Communication Interfaces have been applied as a way to foment social networking and citizen engagement with the community. Shigeno et al. [7] proposed a new model of shared boards, tailored for low-income communities. This technology can be seen as a new type of social media, since it establishes an open digital interface for intra-community message exchange between members and has the potential to foster a new mode of citizen participation.

Koch *et al.* [9] proposed a platform that embeds the concepts of crowd/social computing. The platform orchestrates citizens and sensing cities, interconnecting parties, analyzing and correlating events and providing recommendations and feedback reports. The system, built on the IBM Smarter Cities project was not developed

---

[1] Accessible Way - http://accessibleway.org/.

specifically for the case of blind users. However its outcomes yield significant social contributions. By using it, administrators can make reliable decisions that will impact social services, traffic, energy and utilities, public safety, retail, communications and economic development. Crowd participation can be adjusted to solve problems in the social inclusion and accessibility domain. The major difference is in the services provided. The CanIHelp platform [10] is an embodiment of the concept of inclusive collaboration resulting in an orchestrated model using mechanisms of collective intelligence through social inclusion initiatives. The platform is designed to integrate assistive technologies, collaborative tools and multiple multimedia communication channels, accessible through multimodal interfaces for universal access. The proposed approach combines and generalizes the usage of human computation in a collaborative environment with assistive technologies creating redundancy and complementarity in the solutions provided, contributing to enhance the quality of life of people with special needs and the elderly.

2.2 Blind Users Navigation and Orientation

Location and navigation systems have become widely available in recent years. They are used as a tool for finding a route to a specific destination, to explore the surrounding environment using the provided contextual information or nearby points-of-interest (POI). The SmartVision and Blavigator projects aim to explore information and communication technologies to provide an aid for bind navigation and orientation. The materialization of the project was to develop a cheap and reliable solution for enhancing blind autonomy. This device should be extremely easy to carry and to be used while providing all the necessary help for autonomous navigation [11, 12]. The device was designed to be an extension of the white cane, not a replacement. Another requirement was the usage of ubiquitous technology capable of issuing warning signals when approaching a possible obstacle, a point-of-interest or when the footpath in front is curved and the heading direction should be adapted. The IoT brought new perspectives for the research on systems for search blind navigation and orientation. Ren et al. [13] characterize the IoT as an *"emerging service model [...] forming an unprecedentedly powerful mobile cloud to provide pervasive data collecting, processing, and computing services"*. The overage of sensors arises a new problem linked to the quantity and quality of collected data. New data fusion techniques are needed to ensure the information extraction. The multi-sensor data fusion [14] aggregates data acquired by individual sensors with different characteristics for enhancing efficiency. Akhoundi and Valavi [14] proposed a rule-based fuzzy system for fusion of raw data obtained from the sensors having complement characteristics. Machine-learning techniques are also used to fuse data. Golding and Lesh [16] proposed a system for indoor navigation capable of inferring context-awareness by integrating information from accelerometers, magnetometers, temperature and light sensors. In this area, some of the most important advances are associated with autonomous driving. In the analysis of these systems it is necessary to keep in account that, in the case of the blind the processing shall follow a different treatment since the human factor is crucial in the process of navigation and guidance, as a decision-making system. Therefore the degree of uncertainty is higher than in autonomous driving systems. Crane et al. [20] propose a sensor system with four different sensor types to identify obstacles. A sensor fusion approach that was developed whereby the output of all sensors was in a common grid based format. The system was used for autonomous navigation of an unmanned ground vehicle. Schueler et al. [19] propose an approach for 360-degree multi sensor fusion for static and dynamic obstacles for automotive vehicles. The method combines the advantages of model based object tracking and an occupancy map for the perception of static and dynamic obstacles. The combination of color and infrared (IR) imagery was used in [17] for obstacle detection in autonomous off-road navigation. The authors use data fusion and machine learning for increasing the reliability of the system. The robust detection of obstacles is also addressed by [18], exploiting the environment to predict future behavior of the obstacle and incorporating these hypotheses into the planning process to produce safer actions.

One of the outlooks for inclusion of human perception is using human computation in the process of acquisition, fusion and/or data mining. Mobile crowdsourcing has been gaining momentum as a feasible solution for solving very large-scale problems [13]. Despite the solutions for the key challenges in mobile crowdsourcing, the authors argue that there are still open questions regarding community oriented mobile crowdsourcing and big data applications by mobile crowdsourcing. The Accessibility Social Sensing is a system designed to collect data about urban and architectural accessibility and to provide users with personalized paths, computed on the basis of their preferences and needs. The system combines data obtained by sensing, crowd-sourcing and mashing-up with main geo-referenced social systems, with the aim of offering services based on a detailed and valid data set [15].

An emerging perspective is the combination of emotions/perceptions as a planning method [21]. The major goal is understanding of how people perceive and respond to static and dynamic urban contexts in both time and geographical space. The resulting novel information layer provides an additional, citizen-centric perspective for urban planners.

3   A Model for Enhancing Blind Navigation

In the last decades, addressing the challenging features and requirements of blind navigation has been a research hot topic. The redundancy of the information and location sources using active and passive sensors, the sensing of the user's surroundings using computer vision, the interaction with the user and his/her safety have been some of the prominent themes [9]. Nowadays, in the navigation and orientation tasks one of the key issues is the decision-making capacity, and therefore, these support systems may be considered as decision support systems. As a decision support system, the data available and its accuracy and reliability determines the limit of the system capacity. One perspective into optimizing such a system is through the enhancement of the available data. The use of human computation appears as a possible solution.

An enhanced navigation solution for the blind in a city would aim to: (i) map safe routes for people with accessibility issues (ii) inform about the proximity of points of interest, for example problematic points; (iii) recommend alternative routes when the proximity of trouble spots is detected. Therefore, the solution will evolve the previous developed solution [12] extending it for the blind. The collection and processing of situational data as geo-tagged reports and social aspects is the first part of the process.

The solution works on data collected from multiple users (blind or not) who interact with the system using a mobile application that collects data from the walking behaviour in the sidewalks. The smartphone sensors, such as the accelerometer and gyroscope, capture movement patterns. Variations to the normal moving patterns can represent obstacles, which can be identified by the interaction of users with the device, or by computer vision using a wearable camera. Moreover, when considering public places frequented by hundreds of people, the same information can be grabbed by many users and therefore validated, being added as a temporary obstacle or accessibility barrier to the geographic information system (GIS). The GIS information is used by the navigation system to trace the route to the destination avoiding the barriers, or warning the user about their existence when used in orientation mode. The navigation system would also be able to adapt the route according to notifications received in real time about the obstacles in the path.

The process is summarized in Fig. 1. The user starts the process (1) by collecting information using the sensors of the smartphone. The data collected depends on each device. The use of GPS, accelerometer and gyroscope data is required, at least. If available, other information can be collected as well. This raw information is periodically sent to the server to be processed.

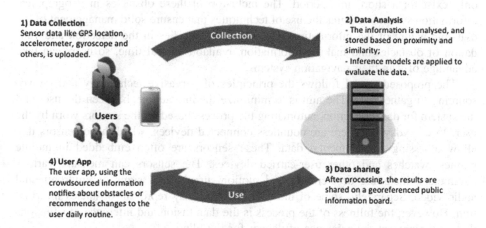

1) Data Collection
Sensor data like GPS location, accelerometer, gyroscope and others, is uploaded.

Collection

Users

2) Data Analysis
- The information is analysed, and stored based on proximity and similarity;
- Inference models are applied to evaluate the data.

4) User App
The user app, using the crowdsourced information notifies about obstacles or recommends changes to the user daily routine.

Use

3) Data sharing
After processing, the results are shared on a georeferenced public information board.

Fig. 1. Proposed high level process architecture

The information is analyzed in order to predict context changes (2), like obstacles or common path changes. This is then weighted by commonality incidence, and shared on a geo-referenced public shared board (3), where users can have access to the information. The user application is based on the user's habits and history, detecting if new events are reported on the server, which can impact the user's daily life and, therefore, notify the user in the most effective way. One example of this situation is when a blind user is walking on a sidewalk under maintenance where the works only started on that day. If the event is detected, and there is information on the server about the works in progress, the application is automatically notified and the route is changed or the user is warned about the hazard.

Emerging technologies will allow evolving and implementing the proposed model. Voice control and monitoring technologies, known as "vox" are evolving, and industry is developing solutions that will be "listening 24/7" harvesting data from what they hear, as well as accepting commands. With the massive use of wearables, Smart TVs and other similar devices, the presence and use of microphones will be ubiquitous: in every room in our homes, in the car, in the office, clipped to a shirt and, of course, in our phones, smart watches, smart glasses and elsewhere. The convergence of ubiquitous computing, wearable computing and, above all, artificial intelligence virtual assistants will make voice-based interface - "hearable computing"[2] - the default way we interact with computers and the Internet.

4 Application Scenario

Cities are filled with obstacles that represent dangers for the blind. In general these obstacles can be classified as static and dynamic.

The static obstacles remain over long periods in the same location, being easily incorporated after their identification into a geographic information system and made available to the navigation systems. Conversely, the dynamic obstacles, or temporary, only exist for a short time period. The inclusion of these obstacles in geographical information systems requires the use of techniques that ensure solid management of the dynamic nature of the information. The key challenge lies in the detection and validation of obstacles and making information available in real time, so users can take advantage of it in their navigation systems.

The proposed model follows the principles of pervasive technology and crowdsourcing to gather data. The aim is to minimize the interactions between the user and the system for data collection, automating the process based on the sensors worn by the user. In everyday life there are countless connected devices, which have sensors that allow assessing environmental data. These sensors are often embedded in mobile phones, watches and other user-carried devices. The sensors can measure various parameters, including acceleration and motion, user's vital signs or environmental audio/video. Separately, some of the data obtained may represent obstacles information. However, the fullness of the process is the data fusion and inference of situations that may represent the existence of danger for the blind.

To illustrate the application of the model, we will follow by demonstrating its use in real cases reflecting an unknown environment (that may be more favorable to the use of the technology) and a known environment.

Consider a familiar environment: as usual each morning, a blind person travels to his/her work. Along the way the sidewalks comply with the regulatory dimensions. However his/her usual route is clogged with a car parked on the sidewalk, leaving a

---

[2] http://www.datamation.com/commentary/listen-up-hearable-computing-is-the-next-big-thing-1.html.

meager space for people to move. The blind detects the obstacle with his/her white cane, forcing him/her to deviate from his/her usual route to get past the obstacle. While surrounding the obstacle the blind gets exalted by boosting the heart rate and saying a few words: "There is no respect. The sidewalks are for people not to park cars". The blind carries a smartwatch with the application for navigation and guidance installed. The application records data collected by the sensors. The data usually collected is the location information associated with acceleration and movement variations (captured by accelerometer and gyroscope), the heart rate and the sounds are captured in these exceptional occasions. The collected data is pre-processed in order to be sent to the host/cloud. This process includes a pre-data fusion, so that it can be combined with the user's history restrained in the device. This intersection lets variations be detected in the usual way of the user and forwards this information to the host/cloud, without identifying the user and ensuring his/her privacy. Furthermore, this information is essential to realize the presence of an anomaly in the usual course and to combine with other sensors the perception of the source/reason for this change. The collected data is sent to the host/cloud where it is fused and extracted the information that is added to the geographic information system. The inclusion of the extracted information depends upon the level of accuracy associated with the information, and according to the model of participation linked to the model.

Now consider the same scenario, but with a person walking on the street and writing an SMS or replying to an email on the smartphone. The information collected in the earlier case when passing the obstacle, is distributed by the system users. Considering that the person is a user of the system and of the device, an installed app recognizes that the user's attention is in writing the message and not in the route (by variations in accelerometer and gyroscope detecting the movement, by recognizing the face looking at the screen through the front camera and the interactions with the application) triggering an alarm indicating danger before crossing the checked area. The alert indicates that there is an obstacle in the street and asks for the attention and collaboration of the user to confirm if it remains at the site, using a simple yes or no question. Upon the user's response, combined with other similar responses, the system is upgraded regarding the accuracy and probability of existence of the obstacle. The retro feedback system allows responding to the dynamics of today's urban environments with constant obstacles that may pose dangers for the blind (and the general population due to the ubiquitous use of technology, deriving in situational induced impairments and disabilities).

In an unknown environment, the blind typically explores the environment in a different way, walking more slowly. Furthermore there is also a greater demand for benchmarks that can be recognized by the system and which can be given additional feedback. When faced with the same obstacle in an unfamiliar environment the reaction of the blind will be less reactive, since he/she was discovering the environment when he/she found the obstacle. Therefore the data collected by the sensors are different. The contextualization of the data in the fusion process is therefore a key factor for ensuring the precision of the information extraction from data collected. Thus, the process once more goes through two phases: one in the device, taking into account the user information so the extraction and selection of data to share can be carried out, keeping the user privacy; and another on the host/cloud.

5 Discussion

The presented conceptual model solves some of the problems associated with the dynamism of the spaces and the dangers it represents for the blind. The model is generic, apart from the capacity and accuracy of each sensor being capable of processing heterogeneous data and fusing it to extract information.

The data collection process is dependent on the development of technologies sensors, and their capacity of communication and interaction with the surrounding environment. Given the advances in the IoT it is expected that within the next years there are several developments in this area, with the emergence of new more accurate wearable devices, such as smartwatches and smartglasses. The independence of the model of data collection ensures its timelessness and adaptation to emerging technologies for data acquisition.

The analysis of the collected data is done in the sensor/device and the host/cloud. The option to phase in the processing of data allows collected data to be crossed with personal and historical data of the user, ensuring the privacy and anonymity. The pre-treatment carried out on the device extracts information from the data taking into account the user's profile, that can then be processed at the cloud/centrally fusing the global with generic users' data. Either data analysis phases are divided into two stages of processing: fusion; and information extraction.

The fusion process categorizes and sorts the data from the various sensors, preparing it for the information extraction process. The separation of this phase allows the modularization of the system and the support for future sensors and devices. For its part, the extraction of information combines the values of each category of sensor, starting with a data consistency assessment. This analysis rules out potential errors that exist in the data captured preventing their spread. Filtering data requires the existence of more than one sample belonging to the same category, as well as reference values and contextual history. It is noteworthy that the process does not eliminate or discard data, only tags it as inconsistent in a given iteration. In the following phases various techniques to extract information are combined. The results involve the extracted information and the accuracy of the process. The availability of this information in a public platform enables its validation and updating by its users at any moment, using the retro-feedback mechanisms.

The retro-feedback mechanisms use two complementary methods of data collection: human computation/crowdsourcing; and sensors/IoT. These mechanisms allow a fast adaptation of the system, reflecting the current dynamics of intelligent environments. However it is pertinent to assess the need for generating data in these environments, so there is a consequent adaptation of the system data. This data capture is facilitated with the use of wearable devices and IoT that open up new possibilities for the application of such models.

6 Conclusions and Future Work

This paper proposes a model for enhancing blind navigation based on the use of crowdsourcing and ubiquitous sensing and computing to generate happy maps that can be dynamically changed with community inputs about dangers in the route.

The proposed approach follows a modular architecture, which interacts with environmental sensors to gather information and process the acquired data with advanced algorithms empowered by human computation.

The use of pervasive technology to acquire environmental data in IoT approach allows users to contribute for its update at each moment, without thereby interfering with their common practices and requiring no interactions. The data collected undergoes dual processing to ensure the privacy and anonymization. It is expected that the aggregation of data collected by multiple users and devices, combined with data fusion techniques ensures high accuracy rates and therefore gaining the trust of users.

Moreover, the gathered metadata will enable the creation of happy maps that are delivered to the blind users through a previously developed navigation system. If the adoption of such a system reaches a level where accessibility issues are reported with high frequency, it may be possible to create a dynamic accessibility map. This type of scenario would also result in an increase of the community engagement on using ubiquitous computing to develop smart cities and extend the support to other communities of users, not limited to visual impairment.

References

1. Quercia, D., Aiello, L.M., Schifanella, R.: The shortest path to happiness: recommending beautiful, quiet, and happy routes in the city. In: Proceedings of Hypertext (2014)
2. Schuurman, D., Baccarne, B., De Marez, L., Mechant, P.: Smart ideas for smart cities: investigating crowdsourcing for generating and selecting ideas for ICT innovation in a city context. J. Theor. Appl. Electron. Commer. Res. 7(3), 49–62 (2012)
3. Crowdsourcing Cooking Book for Cities, Open Cities EU project (2011–2013), Grant Agreement number: 270896, Project Title: OPEN INNOVATION Mechanism in Smart Cities
4. Cardonha, C.H., Gallo, D., Avegliano, P., Herrmann, R., Koch, F.L., Borger, S.: A crowdsourcing platform for the construction of accessibility maps. In: Proceedings of 22nd International World Wide Web Conference (WWW). Rio de Janeiro, Brazil (2013). ISBN 978-1-4503-1844-0
5. Shigeno, K., Borger, S., Gallo, D., Herrmann, R., Molinaro, M., Cardonha, C.H., Koch, F. L., Avegliano, P.: Citizen sensing for collaborative construction of accessibility maps. In: Proceedings of 10th International Cross-Disciplinary Conference on Web Accessibility (W4A 2013) (2013)
6. Fernandes, H., Faria, J., Paredes, H., Barroso, J.: An integrated system for blind day-to-day life autonomy. In: The Proceedings of the 13th International ACM SIGACCESS Conference on Computers and Accessibility (ASSETS 2011), pp. 225–226. ACM, New York, NY, USA (2011)
7. Shigeno, K., Herrmann, R., Mollinaro, M., Cardonha, C., Borger, S., Koch, F.L.: Inferring social intelligence from open communication interfaces. In: Proceedings of 6th International Conference on Social Computing (SOCIOCOM 2014). Academy of Science and Engineering. Stanford, USA (2014). ISBN 978-1-62561-000-3
8. Zhu, T., Xiao, S., Zhang, Q., Gu, Y., Yi, P., Li, Y.: Emergent technologies in big data sensing: a survey. Int. J. Distrib. Sensor Netw. 501, 902982 (2015)

9. Koch, F., Cardonha, C., Gentil, J.M., Borger, S.: A platform for citizen sensing in sentient cities. In: Nin, J., Villatoro, D. (eds.) CitiSens 2012. LNCS, vol. 7685, pp. 57–66. Springer, Heidelberg (2013). ISBN 978-3-642-36073-2
10. Paredes, H., Fernandes, H., Sousa, A., Fortes, R., Koch, F., Filipe, V., Barroso, J.: CanIHelp: a platform for inclusive collaboration, Universal Access in Human-Computer Interaction (2015)
11. Fernandes, H., du Buf, J., Rodrigues, J.M.F., Barroso, J., Paredes, H., Farrajota, M., José, J.: The smartvision navigation prototype for blind users. J. Digit. Content Technol. Appl. 5(5), 351–361 (2011)
12. Fernandes, H., Adão, T., Magalhães, L., Paredes, H., Barroso, J.: Navigation module of blavigator prototype. In: World Automation Congress 2012, Proceedings of the World Automation Congress, Puerto Vallarta (2012)
13. Ren, J., Zhang, Y., Zhang, K., Shen, X.: Exploiting mobile crowdsourcing for pervasive cloud services: challenges and solutions. IEEE Commun. Mag. 53(3), 98–105 (2015)
14. Akhoundi, M.A.A., Valavi, E.: Multi-Sensor Fuzzy Data Fusion Using Sensors with Different Characteristics. arXiv preprint arXiv:1010.6096 (2010)
15. Prandi, C., Salomoni, P., Mirri, S.: mPASS: integrating people sensing and crowdsourcing to map urban accessibility. In: Proceedings of the IEEE International Conference on Consumer Communications and Networking Conference, pp. 10–13 (2014)
16. Golding, A.R., Lesh, N.: Indoor navigation using a diverse set of cheap, wearable sensors. In: The Third International Symposium on Wearable Computers, 1999, Digest of Papers, pp. 29–36. IEEE, October 1999
17. Dima, C.S., Vandapel, N., Hebert, M.: Classifier fusion for outdoor obstacle detection. In: 2004 IEEE International Conference on Robotics and Automation, Proceedings, ICRA 2004, vol. 1, pp. 665–671. IEEE (2004)
18. Ferguson, D., Darms, M., Urmson, C., Kolski, S.: Detection, prediction, and avoidance of dynamic obstacles in urban environments. In: Intelligent Vehicles Symposium, 2008 IEEE, pp. 1149–1154. IEEE (2008)
19. Schueler, K., Weiherer, T., Bouzouraa, E., Hofmann, U.: 360 degree multi sensor fusion for static and dynamic obstacles. In: Intelligent Vehicles Symposium (IV), 2012 IEEE, pp. 692–697. IEEE (2012)
20. Crane III, C.D., Armstrong II, D.G., Ahmed, M., Solanki, S., MacArthur, D., Zawodny, E., Gray, S., Petroff, T., Grifis, M., Evans, C.: Development of an integrated sensor system for obstacle detection and terrain evaluation for application to unmanned ground vehicles. In: Defense and Security, pp. 156–165. International Society for Optics and Photonics (2005)
21. Zeile, P., Resch, B., Exner, J.P., Sagl, G.: Urban emotions: benefits and risks in using human sensory assessment for the extraction of contextual emotion information in urban planning. In: Planning Support Systems and Smart Cities, pp. 209–225. Springer International Publishing, Berlin (2015)

Incorporating Mitigating Circumstances into Reputation Assessment

Simon Miles[1]([✉]) and Nathan Griffiths[2]

[1] King's College London, London, UK
simon.miles@kcl.ac.uk
[2] University of Warwick, Coventry, UK
nathan.griffiths@warwick.ac.uk

Abstract. Reputation enables customers to select between providers, and balance risk against other aspects of service provision. For new providers that have yet to establish a track record, negative ratings can significantly impact on their chances of being selected. Existing work has shown that malicious or inaccurate reviews, and subjective differences, can be accounted for. However, an honest balanced review of service provision may still be an unreliable predictor of future performance if the circumstances differ. Specifically, mitigating circumstances may have affected previous provision. For example, while a delivery service may generally be reliable, a particular delivery may be delayed by unexpected flooding. A common way to ameliorate such effects is by weighting the influence of past events on reputation by their recency. In this paper, we argue that it is more effective to query detailed records of service provision, using patterns that describe the circumstances to determine the significance of previous interactions.

Keywords: Reputation · Trust · Provenance · Circumstances

1 Introduction

In online service-oriented systems, an accurate assessment of reputation is essential for selecting between alternative providers. Existing methods for reputation assessment have focused on coping with malicious or inaccurate ratings, and with subjective differences, and do not consider the full interaction history and context. The context of previous interactions contains information that could be valuable for reputation assessment. For example, there may have been mitigating circumstances for past failures, such as where a freak event affected provision, or a previously unreliable sub-provider has been replaced. Existing methods do not fully take into account the circumstances in which agents have previously acted, meaning that assessments may not reflect the current circumstances, and so be poor predictors of future interactions. In this paper, we present a reputation assessment method based on querying detailed records of service provision, using patterns that describe the circumstances to determine the relevance of

© Springer International Publishing Switzerland 2015
F. Koch et al. (Eds.): CARE-MFSC 2015, CCIS 541, pp. 77–93, 2015.
DOI: 10.1007/978-3-319-24804-2_6

past interactions. Employing a standard provenance model for describing these circumstances, gives a practical means for agents to model, record and query the past. Specifically, the contributions of this paper are as follows.

- A provenance-based approach, with accompanying architecture, to reputation assessment informed by rich information on past service provision.
- Query pattern definitions that characterise common mitigating circumstances and other distinguishing past situations relevant to reputation assessment.
- An extension of an existing reputation assessment algorithm (FIRE [7]) that takes account of the richer information provided in our approach.
- An evaluation of our approach compared to FIRE.

An overview of our approach, with an example circumstance pattern and a high-level evaluation, appears in [10]. This paper extends that work, presenting an in-depth description of the approach and architecture for provenance-based reputation, additional circumstance patterns, and more extensive evaluation.

Reputation and trust are closely related concepts, and there is a lack of consensus in the community regarding the distinction between them [11]. For clarity, in this paper we use the term *reputation* to encompass the concepts variously referred to as trust and reputation in the literature.

We discuss related work in the following section, before presenting our approach in Sect. 3. The baseline reputation model is described in Sect. 4 and we present example circumstance patterns in Sect. 5. Evaluation results are described in Sect. 6 and our conclusions in Sect. 7.

2   Background

Given the importance of reputation in real-world environments, there continues to be active research interest in the area. There are several effective computational reputation models, such as ReGreT [13], FIRE [7], TRAVOS [16] and HABIT [15] that draw on direct and indirect experiences to obtain numerical or probabilistic representations for reputation. In dynamic environments, where social relationships evolve and the population changes, it can be difficult to assess reputation as there may be a lack of evidence [1,7,8,14]. Stereotypes provide a useful bootstrapping mechanism, but there needs to be a sufficient evidence base from which to induce a prediction model [1,3,14,18].

Where there is little data for assessing reputation, individual pieces of evidence can carry great weight and, where negative, may cause a provider rarely to be selected, and never be given the opportunity to build their reputation. While reviewer honesty can be tested from past behaviour and dishonest reviews ignored, it is possible for a review to be accurately negative, because of poor service provision, and still not be an accurate predictor of future behaviour. These are examples of *mitigating circumstances*, where the *context* of service provision rather than an agent's ability meant that it was poorly provided, but that context was temporary. Many approaches use *recency* to ameliorate such effects. However, we argue that recency is a blunt instrument. First, recent provision

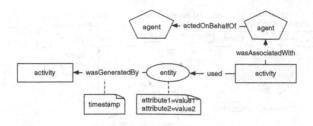

Fig. 1. PROV graph illustrating the key elements

may have been affected by mitigating circumstances, and recency will weight the results higher than older but more accurate data. Second, older interactions may remain good predictors of reliability, because of comparable circumstances.

Instead, we argue for the circumstances of past interactions to be recorded and taken into account more explicitly. This raises the question of what form these records should take, and who should record them. In order to share interaction records between agents, they must be recorded in a commonly interpretable format. PROV is a W3C standard for modelling, serialising and accessing *provenance information*, the history of processes [19]. A PROV document describes in a queryable form the causes and effects within a particular past process of a system (such as agents interacting), as a directed graph with annotations. A visualisation of such a graph, showing PROV's key elements, is shown in Fig. 1. In summary, an *activity* is something that has taken place, making *use* of or *generating entities*, which could be data, physical or other things. *Agents* are parties that were responsible for (*associated with*) activities taking place, and one agent may have been *acting on behalf of* another in this responsibility. Activities, entities and agents (graph nodes) may be annotated with key-value *attributes* describing features that the elements had. *Timestamps* can also be added to show when entities were used or generated by activities.

There has been relatively little use of provenance records for reputation. One of the earliest approaches traversed a decision tree with respect to provenance records to measure reputation [12]. Within the domain of information provision, a richer assessment can be obtained by considering the provenance path of information, the trustworthiness of the information itself, and the reliability of the provider to assess reputation [5,21]. A risk model can be defined that considers the main risk classes and relationships, which can facilitate a detailed risk assessment for an interaction by evaluating the complete provenance path [17].

3   Approach

To enable the use of provenance records to provide personalised reputation assessments, we have proposed the architecture illustrated in Fig. 2, in which clients make requests to an *assessor* for reputation assessments [6]. The assessor relies on provenance graphs to determine reputation, rather than on individual or third party ratings as in existing work. Provenance records are recorded as a

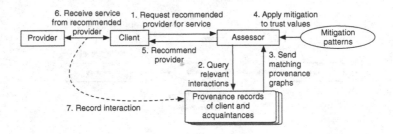

Fig. 2. An architecture for provenance-based service provider reputation

side-effect of interactions, by one or multiple parties, providing crucial evidence that may be missing for assessing reputation [2]. For example, in a logistics chain in addition to clients recording information, providers can record information about sub-contractors, giving information about sub-contractors' performance.

This allows mitigation, situation, indirect responsibility, and other such context to be accounted for, and the interdependencies of providers to be understood. Mitigation can have many forms, such as a subsequently replaced sub-contractor failing to deliver on time, or a client failing to specify required conditions (e.g. expiration date of goods being shipped). The assessor looks for patterns in the provenance that indicate situations relevant to the current client's needs and mitigating circumstances affecting the providers. Provenance data is suitable for this because it includes the causal connections between interactions, and so captures the dependencies between agents' actions. It can include multiple parties to an interaction and their organisational connections. The assessor filters the provenance for key subgraphs from which reputation can be assessed using existing approaches, by identifying successful and failed interactions and adjusting these by mitigation and situation relevance. Assessing reputation in this way avoids the problem of when to update trust, as whenever an assessment is required it is determined using all available evidence.

Reputation enables the assessment and management of the risk associated with interacting with others, and enables agents to balance risk against factors such as cost when considering alternative providers. Such environments can be viewed as service-oriented systems, in which agents provide and consume services. We take an abstract view of service-oriented systems, without prescribing a particular technology. We assume that there are mechanisms for service advertisement and for service discovery. We also assume that service adverts can optionally include details of provision, such as specifying particular sub-providers if appropriate. Finally, we assume that agents record details of their interactions in the form of provenance records, which can be used to assess reputation. The practicality of this last requirement is discussed in Sect. 6.3.

4   Baseline Reputation

Provenance records not only contain rich information that enable reasoning about aspects such as mitigating circumstances, but they also provide a means to maximise the amount of information available for reputation assessment. In this section, we describe how reputation can be driven by provenance records. For the purposes of illustration we consider FIRE [7], but note that other approaches, such as those discussed in Sect. 2 or machine learning techniques, can similarly be adapted to use provenance records.

4.1   The FIRE Reputation Model

FIRE combines four different types of reputation and trust: interaction trust from direct experience, witness reputation from third party reports, role-based trust, and certified reputation based on third-party references [7]. The direct experience and witness reputation components are based on ReGreT [13]. In this paper our focus is on using provenance records of interactions to support reputation, and on defining query patterns for mitigating circumstances. Role-based trust and certified reputation are tangential to this focus, as they are not directly based on interaction records. Therefore, we do not consider role-based trust and certified reputation in this paper (although we do not argue against their usefulness). Reputation is assessed in FIRE from *rating* tuples of the form (a, b, c, i, v), where a and b are agents that participated in interaction i such that a gave b a rating of $v \in [-1, +1]$ for the term c (e.g. reliability, quality, timeliness). A rating of $+1$ is absolutely positive, -1 is absolutely negative, and 0 is neutral. In FIRE, each agent has a history size H and stores the last H ratings it has given in its local database. FIRE gives more weight to recent interactions using a *rating weight function*, ω_K, for each type of reputation, where $K \in \{I, W\}$ representing interaction trust and witness reputation respectively.

The trust value agent a has in b with respect to term c is calculated as the weighted mean of the available ratings:

$$\mathcal{T}_K(a, b, c) = \frac{\sum_{r_i \in \mathcal{R}_K(a,b,c)} \omega_K(r_i) \cdot v_i}{\sum_{r_i \in \mathcal{R}_K(a,b,c)} \omega_K(r_i)} \tag{1}$$

where $\mathcal{R}_K(a, b, c)$ is the set of ratings stored by a regarding b for component K, and v_i is the value of rating r_i.

To determine direct interaction reputation an assessing agent a extracts the set of ratings, $\mathcal{R}_K(a, b, c)$, from its database that have the form $(a, b, c, \_, \_)$ where b is the agent being assessed, c is the term of interest, and "_" matches any value. These ratings are scaled using a *rating recency factor*, λ, in the rating weight function, and combined using Eq. 1. FIRE instantiates the rating weight function for interaction trust as:

$$\omega_I(r_i) = e^{\frac{\Delta t(r_i)}{\lambda}} \tag{2}$$

where $\omega_I(r_i)$ is the weight for rating r_i and $\Delta t(r_i)$ the time since r_i was recorded.

Agents maintain a list of acquaintances, and use these to identify witnesses in order to evaluate witness reputation. Specifically, an evaluator a will ask its acquaintances for ratings of b for term c, who either return a rating or pass on the request to their acquaintances if they have not interacted with b. FIRE uses a variation of Yu and Singh's referral system [22], with parameters to limit the branching factor and referral length to limit the propagation of requests. The ratings obtained from referrals are then used to calculate witness reputation (using Eq. 1, with $\omega_W(r_i) = \omega_I(r_i)$). FIRE assumes that agents are willing to help find witness ratings, and that ratings are honest and credible. In general, these assumptions may not hold and $\omega_W(r_i)$ should account for credibility.

The overall term trust in an agent is calculated as a weighted mean of the component sources:

$$T(a, b, c) = \frac{\sum_{K \in \{I, W\}} \omega_K \cdot T_K(a, b, c)}{\sum_{K \in \{I, W\}} \omega_K} \tag{3}$$

where the reliability of the reputation value for component K is $\rho_K(a, b, c)$, $\omega_K = W_K \cdot \rho_K(a, b, c)$, and W_I and W_W are parameters that determine the importance of each component. The reliability of a reputation value is determined by a combination of the rating reliability and deviation reliability, which characterise a reputation assessment in terms of the number and variability of the ratings on which it is based. The calculations are beyond the scope of this paper (details can be found in [7]), but we note that these metrics can also be calculated from the information in the provenance records.

FIRE does not specify how reputation for different terms is combined into an overall assessment. For simplicity, we assume that terms have equal weight in the same normalised units, and we average across ratings for all terms relevant to a service. Applying varying weights would be a trivial extension.

4.2 Reputation from Provenance Records

As provenance records are not simple tuples containing ratings, unlike in FIRE, we need to determine whether an interaction was good or bad. An interaction's quality could be measured in different terms: the adequacy of the product or service, the speed with which the service was provided, etc. Different terms correspond to different features of provenance graphs. For example, PROV allows timestamps to be added to *use* relations (when an entity began being used by an activity), generation relations (when an entity was generated by an activity), and the start and end of activities. Two timestamps of interest in service provision are the use of the client's request by the service provider, i.e. when the service was requested, and the generation of the service result by the provider, i.e. when the service was completed. Subtracting one from the other gives the duration of service provision. Comparison of this period to the client's expectation gives a rating for the interaction's timeliness term.

Another term could be an observable quality of a product, for example whether a product is damaged. By querying the relevant attribute of the product of a service, a rating can be determined for the quality term. A more interesting term could be the proportion of the product made from materials from sustainable sources. Determining a rating for this latter property would require looking across multiple parts of the provenance graph for an interaction, to determine the sustainability of each component part of the eventual product. For example, to determine the sustainability of a garment details of the fabric and raw materials (e.g. cotton, dye, and fasteners) must also be evaluated. Terms are often domain-specific and are not further discussed here.

5 Circumstance Patterns

PROV data describes past processes as causal graphs, captured from multiple parties and interlinked. The interactions which comprise a service being provided can be described by a sub-graph, and inspecting features of the sub-graphs, such as through a SPARQL query [20], can determine the extent to which they inform reputation. In this section, we specify three mitigating circumstances patterns that could be detected in provenance data. These examples are not intended to be exhaustive, but illustrate the form of such patterns in our approach.

5.1 Unreliable Sub-provider

In the first mitigating circumstance, a provider's poor service on a past occasion was due to reliance on a poor sub-provider for some aspect of the service. If the provider has changed sub-provider, the past interaction should not be considered relevant to their current reputation.[1] This is a richer way of accounting for sub-provider actions than simply discounting based on position in a delegation chain [4]. In other words, Provider A's reputation should account for the fact that previous poor service was due to Provider A relying on Provider B, who they no longer use. The provenance should show:

1. Provider B was used where there was poor service provision,
2. Provider B's activities were the likely cause of the poor provision, and
3. Provider A no longer uses Provider B (not necessarily shown through provenance).

A provenance pattern showing reliance on a sub-provider in a particular instance can be defined as follows. For reference, activities are labelled with An (where n is a number) and entities are labelled with En. Figure 3 illustrates this pattern, along with some of the specific cases below.

Step 1. A client process, A1, sends a request, E1, for a service to a process, A2, for which Provider A is responsible. In the PROV graph, this means that E1 wasGeneratedBy A1, A2 used E1, and A2 wasAssociatedWith Provider A.

Step 2. A2 sends a request, E2, to a service process, A3, for which Provider B is responsible. In the PROV graph, this means that E2 wasGeneratedBy A2, A3 used E2, and A3 wasAssociatedWith Provider B.

Step 3. A3 completes the action and sends a result, E3, back to A2. In the PROV graph, this means that E3 wasGeneratedBy A3, and A2 used E3.

---

[1] Such a situation may indicate poor judgement and so have a degree of relevance, but this is not considered in this paper.

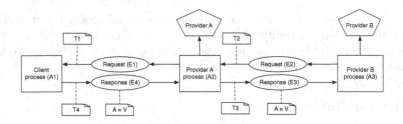

Fig. 3. Provenance graph pattern for unreliable sub-provider circumstance

Step 4. A2 completes the service provision, sending the result, E4, back to A1, so that the client has received the service requested. In the PROV graph, this means that E4 wasGeneratedBy A2, and A1 used E4.

We can distinguish cases in which Provider B would be the likely cause of poor quality service provision. Each case corresponds to an extension of the above provenance pattern.

Case 1. An aspect of the result of provision is poor, and that aspect is apparent in Provider B's contribution. For example, Provider A may have provided a website for a company which appears poor due to low resolution images supplied by Provider B. The extensions to the original pattern are as follows.

- The service provision result, E4, has an attribute $A = V$, which is a reason for the result being poor (e.g. resolution $=$ low).
- The intermediate result from Provider B, E3, has this same attribute $A = V$.

Case 2. The poor provision may not be due to eventual outcome but due to the time taken to provide the service, and this can be shown to be due to the slowness of Provider B. The extensions to the original pattern are as follows.

- The sending of the service request (i.e. the relation E1 wasGeneratedBy A1), is timestamped with T1.
- The receipt of the service result (i.e. the relation A1 used E4), is timestamped with T4.
- The sending of the delegated request (i.e. the relation E2 wasGeneratedBy A2), is timestamped with T2.
- The receipt of the delegated service result (i.e. the relation A2 used E3), is timestamped with T3.
- $T4 - T1 > X$, where X is the reasonable upper limit for the service to be provided, and $T3 - T2 > Y$, where Y is some significant portion of X.

The final criterion required for the above patterns to affect Provider A's reputation assessment is to show that Provider A no longer uses Provider B. This could be through (i) recent provenance of Provider A's provision showing no use of Provider B, or (ii) Provider A's advert for their service specifying which sub-provider they currently use. The latter is assumed the in evaluation below.

We also note that a variation of this pattern is also useful, namely to identify situations in which successful service provision was due to a *good* sub-provider who is no longer used. In this variation the same pattern is used but with poor provision replaced by good provision.

5.2   Freak Event

In the second circumstance, the service provision of Provider A was affected by a one-off substantial event, e.g. ash from an erupting volcano, flooding blocking roads, etc. The freak event can be considered to be an agent in the provenance graph, as it is an autonomously acting entity. The provenance should show:

1. The effects of a known freak event were part of the process of Provider A providing the service, and
2. The part of the process affected by the freak event was the likely cause of the poor service.

The pattern should show that the effects of the freak event were part of the service provision process, illustrated in Fig. 4.

Step 1. A client process, A1, sends a request, E1, for a service to A2 for which Provider B is responsible. In the provenance graph, this means that E1 wasGeneratedBy A1, A2 used E1, and A2 wasAssociatedWith Provider B.

Step 2. A2 begins providing the service by producing entity E2. E2 wasGeneratedBy A2.

Step 3. The relevant effects, A3, of the freak event affect the service provision, so we distinguish what is provided before those effects, E2, and after, E3. A3 used E2, E3 wasGeneratedBy A3, A3 wasAssociatedWith the freak event.

Step 4. The remainder of the service provision process, A4, completes from the state after the freak event has affected the process, E3, and produces the final provision result, E4. A4 used E3, E4 wasGeneratedBy A4.

Step 5. Finally, provision is completed and returned to the client. A1 used E4.

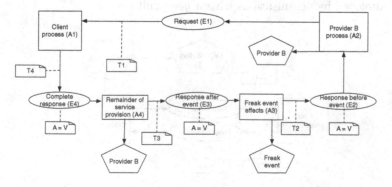

Fig. 4. Provenance graph pattern for freak event circumstance

Similar to the first circumstance above, we can distinguish the cases in which the freak event is the likely cause of eventual poor service. The attributes can indicate that the product before the event (E2) was high quality, while after it (E3) was not, e.g. water damage affecting a parcel. Any delay between the request and response could be primarily due to the freak event (A3).

5.3   Poor Organisation Culture

In the third case, Provider A may be an individual within Organisation B. In such cases, the culture of the organisation affects the individual and the effectiveness of the individual affects the organisation. If Provider A leaves the organisation, this past relationship should be taken into account: Provider A may operate differently in a different organisational culture. The provenance should show:

1. Provider A provided poor service while working for Organisation B, and
2. Provider A is no longer working for Organisation B.

A provenance pattern showing provision of a service within an organisation in a particular instance could be as follows (illustrated in Fig. 5).

Step 1. A client process, A1, sends a request, E1, for a service to A2, for which Provider A is responsible. In the provenance graph, this means that E1 wasGeneratedBy A1, A2 used E1, and A2 wasAssociatedWith Provider A.

Step 2. Provider A is acting on behalf of Organisation B in performing A2. In the provenance graph, this means Provider A actedOnBehalfOf Organisation B in its responsibility for A2 (the latter not depicted Fig. 5 to retain clarity).

Step 3. A2 completes the service provision sending the result, E2, back to A1, so that the client has received the service requested. In the provenance graph, this means that E2 wasGeneratedBy A2, and A1 used E2.

We can then distinguish the cases in which the culture of Organisation B may be a mitigating factor in Provider A's poor provision. Poor performance is identified as described above: either an attribute indicating low quality, a part that is of low quality, or too long a period between the request and response. A variation on the circumstance is to observe where agents were, but are no longer, employed by organisations with a *good* culture.

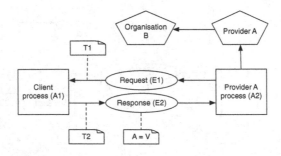

Fig. 5. Provenance graph pattern for poor organisation culture circumstance

6   Evaluation

We evaluated our approach through simulation, comparing it with FIRE, using an environment based on that used in the original evaluation of FIRE [7]. For transparency, the simulation code is published as open source[2].

6.1   Extending FIRE

Existing reputation methods do not account for mitigating circumstances and the context of service provision. The *context* of an interaction is not considered and there is no mechanism for considering mitigating circumstances. In our approach, each agent has its own provenance store, and to determine the reputation of a provider on behalf of a client the assessor queries that client's provenance store and those of its acquaintances. For each interaction recorded in the provenance stores the outcome is considered according to the term(s) that the client is interested in. Since, for illustration, we adopt the FIRE model, the assessor extracts ratings from the provenance of the form $(\_, b, c, i, v)$, where b is the provider in interaction i, and the client in i gave b a rating of v for term c. These ratings are then used to determine reputation (using Eqs. 1 and 3).

Mitigating circumstances and context can be incorporated into existing reputation models by adjusting the weighting that is given to the rating resulting from an interaction for which there are mitigating circumstances. In FIRE, this can be done through the rating weight function, ω_K, for each type of reputation, where $K \in \{I, W\}$, by a factor that accounts for mitigation, specifically:

$$\omega_I(r_i) = \omega_W(r_i) = m \qquad (4)$$

where m is the mitigation weight factor. This factor reflects how convincing an agent considers particular mitigating circumstances, and is defined on a per pattern basis. For the sub-provider and organisation patterns this corresponds to the perceived contribution of a sub-provider or organisation to the service provision, while for a freak event it corresponds to the perceived impact of the event. Mitigation weight factors can be estimated from knowledge of the system and each agent can ascribe a mitigation value to each of its mitigating circumstance patterns. For simplicity, however, we ascribe a global value to each pattern.

Our FIRE implementation calculates trust on the basis of individual and witness experience, i.e. a client's provenance records and those of its acquaintances, applying equal weight to each, but we exclude role-based and certified trust as discussed in Sect. 4. The original evaluation of FIRE allows *exploration* of the space of providers, meaning that the most trusted provider is not always chosen. We include an exploration probability, e, where a client selects the most trusted provider with probability $1 - e$, else will select the next most trusted with probability $1 - e$, etc. This differs from the original evaluation of FIRE which uses Boltzmann exploration to reduce exploration over time. The effectiveness

---

[2] http://bit.ly/1uqLAZO.

of Boltzmann exploration requires the best action to be well separated from others [9]. This is not a reasonable assumption, since providers may be similarly trustworthy. Moreover, there is an assumption that convergence is possible, and in a dynamic environment this is not appropriate.

FIRE's original evaluation divided agents into clients and providers, whereas we assume any agent can be a client or provider. To improve simulation performance we set a memory limit such that, by FIRE's recency weighting, records with a weighting of $\leq 1\%$ are not retained.

6.2   Results

We evaluated the strategies on a simulated network of 100 agents providing services to each other over 1000 rounds. Agents are positioned on, and explore, a spherical world which dictates their neighbours and acquaintances (as in the original evaluation of FIRE [7]), with an average of around 3 neighbours each. This means agents tended to form 2 to 4 clusters of acquaintances. There were 5 primary capabilities (types of service which may require sub-capabilities), capabilities have two terms (quality and timeliness), and each agent has 3 capabilities. Each agent has a 50 % chance to request a service each round and 20 % chance not to pick the most trusted agent. Agents switch sub-provider every 1–15 rounds. Freak events occur with 25 % probability and affected interactions are weighted at 25 % relevancy by our strategy. Where recency scaling was applied, it was set such that after 5 rounds it is 50 % weight. There are 10 organisations, 30 % with a poor culture, reducing the terms of the services provided, while 70 % had a good culture. Agents change organisation every 1–15 rounds. The utility gained in a round is the sum of utility gained per service provision, where the latter is the average of quality and timeliness of the provision (each in $[-1,1]$).

We compared five strategies: FIRE, our approach (Mitigating) with and without recency, FIRE without recency, and random selection. Each strategy was evaluated in 50 networks and the results averaged. Figure 6 shows the results

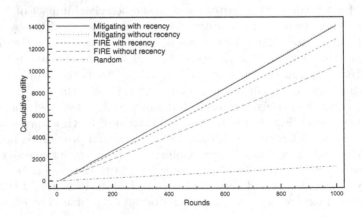

Fig. 6. Cumulative utility over time for all mitigating circumstances patterns.

where all three example circumstances are present (poor sub-providers, freak events, poor organisational culture). Our approach has improved performance, both with and without recency, over FIRE, with an improvement of 10.1 % without and 9.3 % with recency scaling respectively. The recency scaling of FIRE is also shown to be beneficial where mitigating circumstances are not taken into account, i.e. FIRE is better than FIRE without recency. These results match the intuition that recency is valuable for taking account of changes in circumstances, but is crude compared to what is possible when past circumstances are visible. When recency is combined with mitigating circumstances there is negligible improvement, further supporting this intuition.

We also considered how utility varied over a simulation, to better understand the results above. Figure 7 shows the per-round utility for an extract of a single simulation for FIRE and our approach without recency (other approaches are omitted for clarity). Utility varies significantly over time, as changing circumstances mean the most trusted agents may not be the best providers. Our approach has more and higher peaks than FIRE, leading to the higher cumulative utility described above. We believe that this is because our strategy recovers from a change in circumstance more quickly than FIRE. While FIRE's recency scaling means that irrelevant past circumstances are eventually ignored, our approach immediately takes account of the difference in past and present circumstances.

To understand how individual circumstances contributed to the results, we simulated the system with a single circumstance pattern applied. In the case of freak events (Fig. 8a) our approach performs similarly to FIRE, with a small improvement (1.1 % in cumulative utility over 1000 rounds). As expected, FIRE without recency performs worse. Our approach has similar results with and without recency, implying that for a low incidence of freak events (25 %), consideration of recency along with mitigating circumstances has little effect. For unreliable sub-providers (Fig. 8b), there is value to scaling by recency in addition to considering mitigating circumstances. Our approach with recency performs

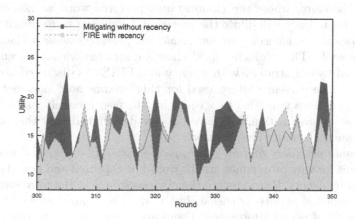

Fig. 7. Per-round utility over one simulation

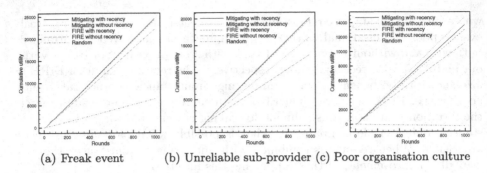

(a) Freak event   (b) Unreliable sub-provider (c) Poor organisation culture

Fig. 8. Cumulative utility for use of the individual mitigating circumstances patterns.

similarly to FIRE (with a 1.6 % improvement), but without recency scaling the utility is significantly lower. Note that both variants of the sub-provider pattern are used, and both poor and good interactions are scaled. With poor organisation culture (Fig. 8c) our approach, with and without recency, outperforms FIRE, with the largest improvement without recency (13.2 %). Here recency scaling reduces performance, and we believe this is because the pattern identifies appropriate situations, and additional scaling reduces the impact of relevant ratings.

6.3 Discussion

In this section, we attempt to answer questions about the results and approach.

Why does accounting for recency seem to be a disadvantage in some results? Recency accounts for changes between the past and present, allowing obsolete information to be forgotten. Weighting relevance by matching against the current circumstance based on provenance patterns aims to account for the past more precisely. Therefore, where the circumstance patterns work as expected, also accounting for recency will dilute the precision, producing worse results.

Why does the result with just unreliable sub-providers show a disadvantage for our approach? The results in Fig. 8b show our strategy without recency being outperformed by our strategy with recency and FIRE. As discussed above, this suggests that the current pattern used for this circumstance does not provide the correct relevance weighting to account for the past precisely, and so recency is a valuable approximation. We have not yet determined why this pattern is imprecise, and it is under investigation.

Why would providers capture provenance graphs? In a practical system, we must account for why provenance graphs would be captured and how they would be accessed by clients. Providers are the obvious source of the provenance data, as it is a record of service provision, but it may be against their interests to release records of poor performance. There are a few answers to this question, though full exploration of the issue is beyond the scope of this paper. First, contractual agreements between clients and providers can require some recording of details as part of providing the service, possibly with involvement of a notary to help ensure validity. In many domains such documentation is a contractual obligation, e.g. journalists must document evidence capture and financial services must document processes for audit. Second, the entities in the provenance graphs are generally exchanged in messages between parties, so there are two agents that can verify the entities were as documented (a commonly used mechanism for non-repudiation). Finally, at a minimum, some information should be present in the client-accessible service advert at the time of service provision, e.g. the organisation to which the provider belongs or sub-provider they use.

What is the value of using PROV graphs over simpler forms? The information recorded in each circumstance (sub-provider, organisation, freak event, etc.) could be provided in a simpler form than a PROV graph, e.g. a tuple. However, a PROV graph is of more practical value. First, every circumstance is different and there may be a varied set of circumstances considered over time, so a single typed tuple is inadequate. Second, the contents of provenance graphs can be collated from data recorded by a set of independent agents, and so it is essential that the provenance follows a standard (W3C PROV). Third, and related, by using PROV there are defined serialisations which mean that clients have a standard means to query the data, e.g. by SPARQL over RDF PROV.

7 Conclusions

In this paper we have described how provenance records can be used to provide the information needed to assess reputation. We have shown how provenance records can be queried to identify when mitigating circumstances occur, to account for context, and argue that this is a more principled approach than simply scaling by recency. Specifically, we defined query patterns for unreliable sub-providers, freak events, and poor organisational culture. The approach is agnostic regarding the reputation model, but for the purposes of evaluation we adopted FIRE [7]. Our evaluation shows that consideration of mitigating circumstances improves performance, but that it is crucial for query patterns to fully capture the context otherwise recency scaling is still required. Future work will define additional query patterns, and develop a method for providing rationale from provenance records explaining reputation assessment.

Acknowledgements. This work was part funded by the UK Engineering and Physical Sciences Research Council as part of the Justified Assessments of Service Provider Reputation project, ref. EP/M012654/1 and EP/M012662/1.

References

1. Burnett, C., Norman, T.J., Sycara, K.: Bootstrapping trust evaluations through stereotypes. In: Proceedings of the 9th International Conference on Autonomous Agents and Multiagent Systems, pp. 241–248 (2010)

2. Burnett, C., Norman, T.J., Sycara, K.: Trust decision-making in multi-agent systems. In: Proceedings of the 22nd International Joint Conference on Artificial Intelligence, pp. 115–120 (2011)
3. Burnett, C., Norman, T.J., Sycara, K., Oren, N.: Supporting trust assessment and decision-making in coalitions. IEEE Intell. Syst. **29**(4), 18–24 (2014)
4. Burnett, C., Oren, N.: Sub-delegation and trust. In: Proceedings of the 11th International Conference on Autonomous Agents and Multiagent Systems, pp. 1359–1360 (2012)
5. Dai, C., Lin, D., Bertino, E., Kantarcioglu, M.: An approach to evaluate data trustworthiness based on data provenance. In: Jonker, W., Petković, M. (eds.) SDM 2008. LNCS, vol. 5159, pp. 82–98. Springer, Heidelberg (2008)
6. Griffiths, N., Miles, S.: An architecture for justified assessments of service provider reputation. In: Proceedings of the 10th IEEE International Conference on e-Business Engineering, pp. 345–352 (2013)
7. Huynh, T.D., Jennings, N.R., Shadbolt, N.R.: An integrated trust and reputation model for open multi-agent systems. J. Auton. Agent. Multi Agent Syst. **13**(2), 119–154 (2006)
8. Jøsang, A., Ismail, R., Boyd, C.: A survey of trust and reputation systems for online service provision. Decis. Support Syst. **43**, 618–644 (2007)
9. Kaelbling, L.P., Littman, M.L., Moore, A.W.: Reinforcement learning: a survey. J. Artif. Intell. Res. **4**, 237–285 (1996)
10. Miles, S., Griffiths, N.: Accounting for circumstances in reputation assessment. In: Proceedings of the 14th International Conference on Autonomous Agents and Multiagent Systems (2015)
11. Pinyol, I., Sabater-Mir, J.: Computational trust and reputation models for open multi-agent systems: a review. Artif. Intell. Rev. **40**, 1–25 (2013)
12. Rajbhandari, S., Contes, A., Rana, O.F., et al.: Trust assessment using provenance in service oriented applications. In: Proceedings of the 10th IEEE International Enterprise Distributed Object Computing Conference Workshops, p. 65 (2006)
13. Sabater, J.: Evaluating the ReGreT system. Appl. Artif. Intell. **18**(9–10), 797–813 (2004)
14. Sensoy, M., Yilmaz, B., Norman, T.J.: STAGE: stereotypical trust assessment through graph extraction. Comput. Intell. (2014)
15. Teacy, W.T.L., Luck, M., Rogers, A., Jennings, N.R.: An efficient and versatile approach to trust and reputation using hierarchical bayesian modelling. Artif. Intell. **193**, 149–185 (2012)
16. Teacy, W.T.L., Patel, J., Jennings, N.R., Luck, M.: Coping with inaccurate reputation sources: experimental analysis of a probabilistic trust model. In: Proceedings of the 4th International Conference on Autonomous Agents and Multiagent Systems, pp. 997–1004 (2005)
17. Townend, P., Webster, C., Venters, C., et al.: Personalised provenance reasoning models and risk assessment in business systems: a case study. In: Proceedings of the 7th IEEE International Symposium on Service Oriented System Engineering, pp. 329–334 (2013)
18. Urbano, J., Roacha, A.P., Oliveira, E.: Refining the trustworthiness assessment of suppliers through extraction of stereotypes. In: Proceedings of the 12th International Conference on Enterprise Information Systems, pp. 85–92 (2010)
19. W3C. PROV model primer (2013). http://www.w3.org/TR/prov-primer/
20. W3C. Sparql 1.1 overview (2013). http://www.w3.org/TR/sparql11-overview/

21. Wang, X., Govindan, K., Mohapatra, P.: Provenance-based information trust-worthiness evaluation in multi-hop networks. In: Proceedings of the IEEE Global Telecommunications Conference, pp. 1–5 (2010)
22. Yu, B., Singh, M.P.: Searching social networks. In: Proceedings of the 2nd International Joint Conference on Autonomous Agents and Multi Agent Systems, pp. 65–72 (2003)

Agent Protocols for Social Computation

Michael Rovatsos[⊠], Dimitrios Diochnos, and Matei Craciun

School of Informatics, The University of Edinburgh, Edinburgh EH8 9AB, UK
{mrovatso,ddiochno,s1374265}@inf.ed.ac.uk

Abstract. Despite the fact that social computation systems involve interaction mechanisms that closely resemble well-known models of agent coordination, current applications in this area make little or no use of the techniques the agent-based systems literature has to offer. In order to bridge this gap, this paper proposes a data-driven method for defining and deploying agent interaction protocols that is entirely based on using the standard architecture of the World Wide Web. This obviates the need of bespoke message passing mechanisms and agent platforms, thereby facilitating the use of agent coordination principles in standard Web-based applications. We describe a prototypical implementation of the architecture and experimental results that prove it can deliver the scalability and robustness required of modern social computation applications while maintaining the expressiveness and versatility of agent interaction protocols.

Keywords: Agent communication · Social computation · Web agents

1 Introduction

Most real-world *social computation* applications that involve large-scale human and machine collaboration (e.g. collective intelligence [10] or human computation [8]), are currently implemented using either *ad hoc* methods or programming frameworks [1,9] that make no use of agent technology. Within the agents community, on the other hand, agent communication languages and interaction protocols [3] have been widely used to design and deploy a wide range of agent coordination mechanisms, many of which bear close similarity to those needed in social computation systems. This is at least in part due to the fact that the *architectural* proposals for developing real-world agent-based systems mostly rely on bespoke platforms with custom message passing mechanisms and control structures. Since the inception of those agent platforms, the architecture of the Web [5] has given rise to a plethora of massive-scale distributed applications, almost in complete ignorance of agent-based techniques [13].

The work presented in this paper aims to bridge the gap between agent coordination techniques and social computation by providing a method for mapping the principles of agent protocol design to the architecture of the Web. We describe a data-driven method for defining and deploying agent interaction protocols that complies with the architecture of the Web, and does away with a

© Springer International Publishing Switzerland 2015
F. Koch et al. (Eds.): CARE-MFSC 2015, CCIS 541, pp. 94–111, 2015.
DOI: 10.1007/978-3-319-24804-2_7

need for point-to-point messaging infrastructures. Also, contrary to many existing agent platforms, it does not assume ideal conditions regarding liveness of agent processes and availability of perfect communication channels.

The basic principles of this architecture are simple: It conceives of messages as entries in persistent data stores accessible via normal HTTP operations, and models dependencies between these messages through an explicit graph structure, where causal and temporal links between messages in a protocol are exposed to agents via Web APIs. This enables avoiding redundant messaging in broadcast situations, failure recovery and management of "stale" interactions, lightweight *ex post* modification of previous interactions, as well as global monitoring and analysis of coordination processes. Also, it leverages the architecture of the Web to enable lightweight communication and is oblivious to the degree of centralisation applied in systems design. We describe a prototypical implementation of our architecture in a typical application scenario that allows us to demonstrate its benefits. Our experiments with a deployed prototype show that our approach offers significant advantages in terms of scalability and robustness.

The remainder of the paper is structured as follows: We start by introducing an example scenario in Sect. 2 that serves to illustrate our framework, and is also later used in our experiments. Section. 3 introduces our formal framework for modelling conventional agent interaction protocols and their semantics. Our data-driven architecture is presented in Sect. 4 together with a discussion of its properties. Experiments are presented in Sect. 5, and after reviewing related work in Sects. 6 and 7 concludes.

2 Example

A typical social computation scenario that involves large-scale agent collectives, and which we will use for illustration purposes throughout the paper, is ridesharing (see, for example, blablacar.com and liftshare.com), where travellers (drivers and passengers) request rides posting location, price, and possibly other constraints. Ridesharing is a representative example both due to the range of functions it requires (matchmaking, negotiation, teamwork) and because it exhibits many characteristics of real-world collective collaboration systems (many users, asynchronous communication, heterogeneity of user platforms).

The *team task* protocol shown in Fig. 1 describes a possible coordination mechanism that could be used in such a system, following a traditional agent-based model which involves an orchestrator o and task peers p in an 1:n relationship. In the top section of the diagram, peers ADVERTISE their capability to play role r in action a, e.g. driving a car, occupying a passenger seat, or paying a driver in the case of ridesharing. This advertisement is acknowledged simply to terminate this stage with a definite response. In the subsequent *matchmaking* stage, peers may REQUEST a task, i.e. a plan that will achieve getting from initial state I to goal state G, subject to certain constraints C (e.g. a price limit).

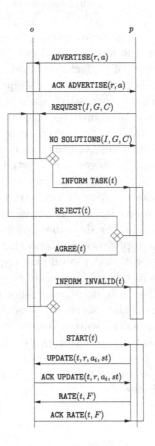

Fig. 1. The team task protocol

Based on requests from various agents and by using the capabilities they have advertised, o proposes a possible task t that would involve a specific plan π to be executed, and a role assignment for the participants clarifying which agent has to perform which actions, or tells p that NO_SOLUTIONS can be found. In the case of ridesharing, the plan would be the ride specification, quoting a price, time, and possibly other constraints (e.g. whether smoking is allowed). If a peer AGREEs to a task, this might have become invalid in the meantime because others have REJECTed it. If the task is still valid, and once all participants agree, the orchestrator invites participants to START executing the plan, after which p can UPDATE the execution status st of individual steps a_i in the plan (e.g. "we reached the destination") or provide feedback reports F regarding the task (e.g. "the driver was driving too fast"). These steps can be repeated depending on how many steps there are in t in the case of UPDATE, or without limitation in the case of RATE. If p rejects a task, more tasks might be suggested until no more solutions exist. Note that the protocol deliberately contains a few "imperfections": First of all, once a peer agrees, she will not be notified if other participants reject the task she agreed too. Secondly, there is no timeout for the negotiation. Hence, a peer will never be told that the negotiation failed because some participants did not respond. Below, we will explain how our proposed architecture results helps address such issues without complete protocol re-design.

3   Formal Framework

Before proposing our own approach, we introduce a formal framework that allows us to capture action and communication semantics in a decentralised agent-based system. Our formalism is based on a plan-based semantics of communication and action, i.e. we consider a state transition system where messages and actions (we use this term to denote non-communicative actions that change the environment state) modify the values of state variables. While this follows traditional STRIPS-style planning formalisms [6], note we are not assuming the use of actual planning algorithms in the system. The notation just gives us a simple, generic way of describing a discrete, distributed state transition system.

Let $V = \{v_1, v_2, \ldots, v_k\}$ variables that range over a domain of discourse D^1, and *constraints* $c = \{(v_1, D_1), \ldots, (v_m, D_m)\}$ denoting that $v_i \in D_i \subseteq D$, where all v_i are distinct, and $\{v_1, \ldots, v_m\} \subseteq V$ is called the *domain dom(c)* of constraint c. We call a constraint *failed* if any for any $v_i \in dom(c)$ we have $D_i = \emptyset$, and write $c = \bot$ in this case. A *substitution* θ is a constraint $\{(v_1, E_1), \ldots, (v_l, E_l)\}$ that can be applied to c to result in a new constraint $c\theta = \{(v_1, D'_1), \ldots, (v_m, D'_m)\}$ such that for all $v_i = v_j$ with $v_i \in dom(c)$ and $v_j \in dom(\theta)$ we have $D'_i = D_i \cap E_j$. A substitution is called a *grounding* if $|D'_1| = |D'_2| = \ldots = |D'_m| = 1$. A grounding is *admissible* if $c\theta \neq \bot$, and we write $\lfloor c \rfloor$ for the set of all possible *groundings* or *instances* of c. Entailment among two constraints is defined as $c \models c'$ if $\lfloor c \rfloor \subseteq \lfloor c' \rfloor$, i.e. c is a stricter constraint satisfied by some groundings of c'.

Next, we introduce agents and their actions. Assume *agents* $A = \{a_1, a_2, \ldots a_n\}$, and, in slight abuse of notation, let their names also be valid variable values, i.e. $A \subseteq D$. We consider a timed system with execution steps $T = \{t_1, t_2, \ldots\}$ using a global clock shared by all agents. For any variable $v \in V$, a_i may have a local copy whose value can change over time. We write v_i^j for the value of v for agent i at time step j (V_i denotes agent i's local variables, and V_i^j their values at time j). We may drop the subscript and write "$v = d$" for some variables whenever all agents' local copies agree on the variable value, i.e. $v = d \Leftrightarrow \forall i . v_i = d$. *Fluents* $F = \{f_1, \ldots, f_k\} \subseteq V$ are variables that describe system states, and exclude any auxiliary variables not used to reflect system state, e.g. the roles denoting senders and receivers of messages in message schemata (see below). A *state specification* S is a constraint with $dom(S) \subseteq F$, and is used to represent the set of all states $s \in S$ with $s \models S$. A state s can be viewed as a constraint that is a full variable assignment $\{(f_1, \{d_1\}), \ldots, (f_k, \{d_k\})\}$ for all

---

[1] Different types D_j can be used here to accommodate different types of variables. These are omitted for simplicity.

domain fluents in F. When referring to states, we will write s_i^j to denote a full assignment to concrete values for agent i at time j.

An *action* $ac = \langle \{a_1, \ldots, a_k\}, pre, eff \rangle$ is performed by agents $\{a_1, \ldots, a_k\} \subseteq A$ and is associated with two constraints, its *preconditions pre* and *effects eff* with $dom(pre) \cup dom(eff) \subseteq F$. For any $s \in S$ with $s \models pre(ac)$ (ac is *applicable* in s), execution of ac results in a *successor state* $succ(s, ac) = s'$ where $s' = s \setminus \{(v, D) | v \in dom(eff(ac))\} \cup eff(ac)$. In other words, if ac is applicable in s, then the successor state results from removing the values of all *affected* variables in $dom(eff(ac))$ from s and adding their new values as per $eff(ac)$. Note that these actions need not be "ground" in the sense that the fluents they involve need to have specific single values before or after the action. A *plan* $\pi = \langle ac_1, \ldots, ac_n \rangle$ is a sequence of actions such that ac_1 is applicable in the states represented by the initial state specification I, i.e. $I \models pre(ac_1)$, and each ac_i is applicable in $succ(s, \langle ac_1, ac_2, \ldots, ac_{i-1} \rangle)$ (where $succ$ is canonically extended to sequences of actions), for all $2 \leq i \leq n$ and $s \models I$. Plans provide the definition for any well-defined sequence of actions that is feasible given the specifications of these actions and the current system state. A plan π is a *solution* for a planning problem $\langle I, G, Ac \rangle$ with initial state specification I and goal state specification G if $succ(s, \pi) \models G$ for all $s \models I$, i.e. if its execution from any state that satisfies I results in a state that satisfies G.

Given this general framework, we can proceed to defining the structure and semantics of agent protocols.

Definition 1. *A* message schema *$\mu =$MSG(se, re, c, pre, eff) is a structure with label* MSG, *where se and re are variables for the sender(s) and receiver(s) of the message, constraint c denotes the message content, and precondition/effect constraints pre and eff with $dom(pre) \subseteq V$ and $dom(eff) \subseteq V$.*

In Fig. 1, such schemata label the edges connecting the individual boxes on the swimlanes of the diagram (which represent sender p and receiver o), e.g. REQUEST$(p, o, \{(i_p, \{I\}), (g_p, \{G\}), (c_p, \{C\})\})$. For readability, we omit preconditions and effects and the constraint notation assigning concrete values to p's local variables for I, G, and C is not used in the diagram (the request implies, for example, that p's local variable i_p has value I at the time of sending).

To define the structure of a protocol, we introduce a graph that is "dual" to that in the diagram, in that is has message schemata for nodes and edges for decision points:

Definition 2. *A* protocol graph *is a directed graph $P = \langle \Phi, \Delta \rangle$ whose node set includes a set of message schemata uniquely identified by message labels, with additional root and sink nodes start and end ($eff(start) = pre(end) = \emptyset$). Its edges are given by a mapping $\Delta : \Phi \rightarrow 2^\Phi$. Every edge (μ, μ') with $\mu' \in \Delta(\mu)$ is labelled with $eff(\mu)$ and $pre(\mu')$.*[2]

---

[2] Throughout the paper, we adopt the convention of referring to elements in a structure $x = \langle y, y', \ldots \rangle$ as $y(x), y'(x)$ etc.

We present the protocol graph for part of the team task protocol (preconditions end effects are only shown for REQUEST):

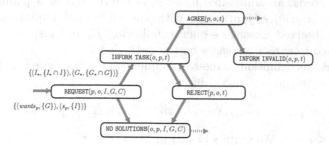

The example assumes that the precondition for REQUEST is $\{(wants_p, G), (s_p, I)\}$ for p. The effect for o, who is gathering planning problems from peers to propose a joint plan that solves all of them, is $\{(I_*, I_* \cap I), (G_*, G_* \cap G)\}$ where "$*$" denotes the view o has of all peers. In other words, o's strategy involves conjunctively narrowing down initial and goal states before suggesting a plan that satisfies all of them. To make the plan-based semantics of protocols concrete, we need to introduce messages as instances of schemata:

Definition 3. *A message is a structure $m = \langle \mu, Se, Re, \theta, t \rangle$ where μ is a message schema, $Se \subseteq A$ and $Re \subseteq A$ are the (non-empty) sets of senders[3]/receivers of the message, θ is a substitution for $c(\mu)$, and t the time the message was sent.*

The following definition defines when a message is *admissible*, i.e. it is a legal continuation of an observed interaction:

Definition 4. *For any protocol graph P, state $s^t \in S$, and initial message sequence $\pi = \langle start, m_1, \ldots, m_{t-1} \rangle$, define:*

$$\langle \pi, s_t \rangle \models_P m_t \quad :\Leftrightarrow \quad m_t = end \wedge end \in \Delta(\mu(m_{t-1})) \quad \vee$$
$$(\theta = \theta(m_1)\theta(m_2)\cdots\theta(m_t) \neq \bot \wedge \mu(m_t) \in \Delta(\mu(m_{t-1})) \wedge$$
$$\forall i \in Se(m_t).\ s_i^t \models pre(\mu(m_t)\theta) \wedge$$
$$\forall j \in Re(m_t).\ s_j^{t+1} \models succ(s_j^t, \mu(m_t)\theta))$$

This defines a message m_t as admissible in the context of a current message sequence and state $\langle \pi, s^t \rangle$, which we call a *model* for m, if either $m_t = end$ and its immediate predecessor was connected to the *end* node in P, or if (i) its schema $\mu(m_t)$ is a successor to that of the most recent message, (ii) the preconditions (effects) of that schema are satisfied by all senders (receivers) of the message in timestep t $(t + 1)$, and (iii) this is subject to the combined substitution θ that

---

[3] Allowing many senders in messages may seem counter-intuitive at first, but is useful for situations where a physical sender acts on behalf of a whole group, or to summarise identical messages received from various peers as one message in the data-driven model we introduce in Sect. 4.

accumulates all the substitutions applied in previous messages (and which must itself be consistent).[4]

In other words, an admissible message is interpreted as a planning action $\langle Se \cup Re, pre(\mu(m)\theta), eff(\mu(m)\theta) \rangle$, with the additional requirement that it extends the observed message sequence following P, and respects the substitutions applied to earlier messages on the path that led to it.

To extend this definition to message sequences, we can write $\langle \pi, s \rangle \models_P \pi'$ for any finite $\pi' = \langle m_{t+1}, \ldots, m_{t+k} \rangle$ iff

$$\langle \pi \langle m_{t+1}, \ldots, m_{t+j} \rangle, succ(s, \langle m_{t+1}, \ldots, m_{t+j} \rangle) \rangle \models_P m_{t+j+1}$$

for all $0 \leq j \leq k - 1$. We write $s \models_P \pi)$ iff $\langle \langle \rangle, s \rangle \models_P \pi$.

With this, we can proceed to define the semantics of a protocol through the set of admissible *continuations* it gives rise to in a specific state given an observed execution history:

Definition 5. *Let $s \in S$ and π a message sequence. If $s \models_P \pi$, the continuations $[\![\pi]\!]_s$ of π are defined as the (potentially infinite) set of sequences messages π' for which $\langle \pi, s \rangle \models \pi'$ holds. We let $[\![\pi]\!]_s := \perp$ if $s \not\models_P \pi$.*

This completes our account of a simple and fairly generic plan-based semantics for agent interaction protocols. Our semantics does not make any specific commitment as to the actual semantic language (e.g. mentalistic, commitment-based, or deontic) used to specify constraints governing the exchange of messages. Instead, it specifies what message sequences are admissible under a shared protocol definition, and how message passing results in a synchronisation among agents' local variables. For simplicity, we have assumed that no additional agent actions or exogenous events occur during protocol execution. Note, however, that such actions or events could be easily accommodated in the protocol graph as additional choices between successive messages without requiring additional formal machinery.

4   Data-Centric Architecture

4.1   Framework

While conventional specifications of agent interaction protocols such as the ones considered above provide a very flexible framework for coordinating multiple agents, the point-to-point message passing they assume can be problematic in large-scale multiagent systems using potentially unreliable communication infrastructures, and operating over long periods of time, so that the contributions of agents occur at unpredictable points in time.

---

[4] Note that different semantics are possible here, which may assume that senders also have a modified state regarding their perception of receivers' local variables after sending a message, or receivers inferring facts about senders' previous states upon receipt of a message. Which of these variants is chosen is not essential for the material provided below.

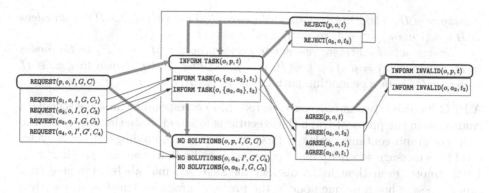

Fig. 2. Data-centric model of part of the ridesharing protocol

Consider a real-world deployment of the protocol shown in Fig. 1 in a web-based ridesharing application with many users. If we use conventional message passing, this protocol would require n conversations for n task peers going on in parallel, and o would need to maintain separate internal data structures to track which agents have already agreed to the task, which of them may provide execution updates, etc. Also, these conversations would have to remain "open" indefinitely, unless strict time limits were imposed on these parts of the protocol. Another drawback is that many data objects such as identical requests, suggested tasks, or information about invalid/agreed tasks and initiation of task execution would have to be sent repeatedly from/to different peers. Finally, if we wanted to de-couple different parts of the protocol that are not causally linked to each other in order to allow for a more flexible execution of the different stages of the protocol (e.g. advertising capabilities is unrelated to negotiation), this would involve creating separate protocols, and managing synchronisation among variables that are not local to a single protocol.

Before introducing our data-driven architecture to address some of these issues, we present its instantiation for the negotiation part of our team task protocol as an example in Fig. 2. The diagram combines the original protocol graph (message schemata in rounded boxes, connected with bold grey arrows) with *message stores* attached to every schema. These message stores contain messages exchanged by the participants so far, and links (black arrows) between messages that were generated in response to each other. As before, we omit preconditions and effects as well as timestep labels and the details of content constraints for readability.

Using linked message stores enables us to replace message passing among agents by inspecting and modifying the contents of persistent message repositories, which is the key idea behind our approach. We start by introducing *protocol execution graphs (PEGs)*, which provide the link structure arising from observed message sequences:

Definition 6. *Let* $\Pi = \{\pi_1, \ldots, \pi_k\}$ *a set of protocol executions where* $\pi_i = \langle m_1, \ldots, m_j, \ldots, m_{t_i} \rangle$ *and* $\pi_{ij} = m_j$, *and* $M(\Pi) = \{\pi_{ij} | \pi_i \in \Pi\}$ *the set of all*

messages in Π. The PEG *is a directed graph* $P(\Pi) = \langle M(\Pi), \Delta(\Pi) \rangle$ *with edges* $\Delta(\Pi) = \{(m, m') | \exists \pi_i \in \Pi.m = \pi_{ij} \wedge m' = \pi_{ij+1}\}$.

For any set of messages, we define a mapping $\varphi : M \to \Phi(P)$ *to the nodes in* P, *where* $\varphi(m) = \mu$ *if* $\exists \mu \in \Phi(P).\mu = \mu(m)$ *and* $\perp$ *else. Given this,* $\pi_i \in \Pi$ *is associated with a generating path* $\varphi(\pi_i) := \langle \varphi(m_1), \ldots, \varphi(m_{t_i}) \rangle$ *in* P.

A PEG has every two messages connected that correspond to message schemata connected in the protocol graph the executions followed. Note that whenever the protocol graph contains cycles, a PEG may contain unfoldings of these cycles (and thus message schemata may appear repeatedly in a generating path $\varphi(\pi_i)$). Furthermore, even though the distinct message schema labels guarantee that every message has a unique node in the protocol graph assigned to it, identical messages (sent to or from different agents) appear only once in the graph. On the other hand, if two messages have identical senders, receivers, and conent, they would count as different nodes in the PEG if they were part of different conversations (as they are annotated with different timestamps). In Fig. 2, the nodes of the PEG are the entries of the boxes under each message schema, and its edges are depicted as black arrows connecting these nodes.

As concerns continuations, we can extend our previous definitions canonically to sets by letting $\langle \Pi, s \rangle \models_P m$ iff $\exists \pi \in \Pi.\langle \pi, s \rangle \models_P m$ and $[\![\Pi]\!]_s := \cup_{\pi \in \Pi} [\![\pi]\!]_s$.

The final step in our construction is to identify *message stores*, one for each message schema μ appearing in the protocol graph (shown as square boxes in Fig. 2). These provide a somewhat orthogonal view of the PEG, focusing on specific message schemata:

Definition 7. *A* message store *is a set of messages* $M_\mu := \{m \in M(\Pi) | \mu(m) = \mu\}$ *containing all message instances for a message schema* μ. *It supports the following operations given* $m = \langle \mu, Se, Re, \theta, t \rangle$:

- $get(a, M_\mu) = \{m \in M_\mu | a \in Re(m)\}$
- $add(a, M_\mu, m) = M_\mu \Leftrightarrow a \in Se(m) \wedge M'_\mu = M_\mu \cup \{m\}$
- $del(a, M_\mu, m) = M'_\mu \Leftrightarrow a \in Se(m) \wedge M'_\mu = M_\mu \backslash \{m\}$
- $mod(a, M_\mu, m, m') = add(a, del(a, M_\mu, m), m')$

The operations add, del (and mod) leave M_μ *unchanged if their arguments do not satisfy the above constraints.*

The main reason we define message stores as first-order citizens in our architecture is that they permit the definition of operations which can be used to emulate sending and receiving messages. These operations, which are realised as physical messages over the network (but we distinguish from protocol messages) allow an agent to create a new message if it is a sender of that message (aff), and to inspect those messages in a store that have her as receiver (get). We also permit deletion of previous messages through del for reasons that will become clear below, and modification of an existing message through mod (a combination of del and add).

Using these methods, a message such as $\text{REQUEST}(a_1, o, I, G, C_2)$ in Fig. 2 would be realised as a sequence of calls $add(a_1, M_\mu, \text{REQUEST}(a_1, o, I, G, C_2)) \to$

$get(o, M_\mu)$ where M_μ is the message store for REQUEST. This enables a different way of processing the protocol specification, which is based on an ability to generate responses to any message contained in a message store without requiring a control flow that manages every conversation sequence individually:

Proposition 1. *Let $s^t \in S$ and $m = \langle \mu, Se, Re, \theta, t \rangle$ a message with $\langle \Pi, s^t \rangle \models_P$ m. We define*

$$op = get(Re, add(a, M(\mu), m)))$$

where $a \in Se$ and $get(Re, \ldots)$ is shorthand notation for all receivers executing the get operation in any ordering. We assume that each get/add operation takes one timestep. Further, we assume that the add operation is only performed if $s_i^t \models pre(\mu\theta)$ for all $a_i \in Se$, and all $a_j \in Re$ update their local state s_j^t to $s_j^{t+1} = succ(s_j^t, \mu\theta)$ instantly when they observe any new message m.

Then, if $|Re| = k$, and no other actions or message store operations are executed between t and $t + k$, it holds that $M_\mu' = op(M) = M \cup \{m\}$ in s^{t+k} and $succ^{(k)}(s^t, op) = succ(s^t, m)$.[5]

PROOF. The proof for this proposition requires only straightforward application of the respective definitions. The operation op on M_μ involves one sender adding m to the message store (which implies $M_\mu' = op(M) = M \cup \{m\}$), and k receivers Re getting the result. Since the message is admissible, we would have $s_i^t \models pre(\mu\theta)$ for all $i \in Se$ and $s_j^{t+1} \models succ(s^t, \mu\theta)$ if this message was sent. We assume that the get message is only sent if the sender can locally satisfy the preconditions of m, and that receivers incorporate the effects of any new message observed on a message store locally (though for a given agent this will only happen at s^{t+l} for some $1 \le l < k$ depending on when the receiver performs the get operation). Given this, and under the assumption that no other action occurs while op is being executed, we have $succ^{(k)}(s^t, op) = succ(s^t, m)$. ∎

The importance of this proposition is twofold: Firstly, it shows how message store operations can correctly replace any protocol message exchange. Secondly, it reveals that an additional $|Re|$ get operations are necessary to produce the same outcome, and that the receivers monitor the contents of each relevant message store continually. On the other hand, it is sufficient if the time k required for these updates is less than the time that passes until further messages being sent to or from the recipients, or other actions are executed that affect their local state. Our model also allows for more unusual operations on message stores, for example deletions of past messages. While this might seem counterintuitive, we discuss in Sect. 4.2 how it can be very useful in real applications. Deletions require a more complex "rollback", which obviously cannot undo the global state of the system, but for which we can establish a weaker result:

Proposition 2. *For any message m, let $next(m, \Pi) = \{m' | (m, m') \in \Delta(\Pi)\}$ with $next^*(m, \Pi)$ as its reflexive and transitive closure. Removing $next^*(m, \Pi)$*

---

[5] The superscript (k) is added to the $succ$ function here to indicate that op requires this number of timesteps.

results in a PEG $\Pi' = \langle \Phi', \Delta' \rangle$ *where* $\Phi' = \Pi \backslash next^*(m, \Pi)$ *and* $\Delta' = \Delta(\Pi)$ $\backslash \{(m', m'') | \{m', m''\} \cap next^*(m, \Pi) \neq \emptyset\}$. *It holds that:*

1. *If* $\forall m' \in next^*(m, \Pi)$. $M'_{\mu(m')} = del(a, M_{\mu(m')}, m_i)$ *and* $\forall m' \notin next^*(m, \Pi)$. $M'_{\mu(m')} = M_\mu(m')$, *then* $M'_\mu = M_\mu(\Pi')$ *for all* μ *and some* $a \in A$.
2. *For any* $m \in \Pi'$ *we have* $\langle \Pi^{t(m)}, s^{t(m)} \rangle \models_P m_j$ *where* $\Pi^{t(m)}$ *and* $s^{t(m)}$ *are the contents of the original PEG and state at time* $t(m)$ *when the message was created.*

Proof. Statement 1. claims that deleting all successors of m from the respective message stores, and leaving all other message stores unchanged will restore the property that any message store M_μ in the system contains all messages instantiating a schema μ in Π'. This is trivially the case, as Π' is identical to Π with the exception of having m and all its successors and their adjacent edges removed. To see that statement 2. holds, it suffices to observe that all remaining messages in Π' are either a predecessor of m, or occur on paths that do not contain m or any of its successors. It follows that their validity at the time of their creation is maintained if we remove m and all subsequent messages. ∎

The main implication of this result is that when a message is deleted from a store, then all its successors need to be deleted with it to maintain some level of consistency (this also assumes that no other messages or modifications on message stores take place in the meantime). Even with these provisions, the level of consistency achieved is obviously much weaker than what can be guaranteed for *add* operations, as deletions remove paths that were previously available, and only paths unaffected by the removal of m have identical continuations as before the removal. Also, the system state may have changed compared to when the original messages were sent, so that we may not be able to track what interactions brought it about. Finally, we should note that the two properties we have just established apply to *mod* operations as a consequence of those operations being abbreviations for a composition of *del* and *add* calls.

4.2   Discussion

To illustrate the use of our model, let us revisit the example from Fig. 2 in more detail: We have six initial REQUEST messages from agents $a_1, \ldots a_4$, which result in possible tasks t_1 for $\{a_1, a_2\}$, t_2 for $\{a_2, a_3\}$ and no solution for a_4 (maybe because his requirements don't match those of any other peers). One immediately obvious advantage of our approach here is that only INFORM_TASK messages need to be "sent" to two agents each (sets $\{a_1, a_2\}$ and $\{a_2, a_3\}$).

Next, we have the situation that a_2 AGREEs to t_2 and a_3 REJECTs this task. We assume that a_2 is the driver and needs to agree first (no ride can be taken without a car), but there is no such restriction regarding rejection, which any participant of the task can issue at any point. Now if a_3 issues the rejection first, a_2 will receive an INFORM_INVALID response, as shown in the diagram, and no agreement on t_2 will be possible anymore. If a_2 has already agreed, however, this agent will never be notified of a_3's rejection, a problem we already mentioned in Sect. 2.

One solution to this problem would be to add an edge from INFORM_INVALID to AGREE, which was not included in the original protocol of Fig. 1. Since previous INFORM_TASK messages also gave a_2 the option t_1, she can now agree to this task, and after a_1 agrees, too, the next message would be START to initiate task execution.

This is generally how protocol flexibility has to be accommodated in normal agent protocols – every possible agent behaviour has to be accounted for by providing additional paths that enable other agents to respond appropriately. In fact, the INFORM_INVALID→AGREE edge would not work here, as no alternative possible task t_1 would be known to a_2 (unless a list of all possible tasks was sent with INFORM_TASK from the outset, which would doubtlessly complicate the workflow further). So, we would have to backtrack at least to the level of INFORM_TASK (in a "task no longer available, here's another alternative" fashion) to allow a_2 to make alternative choices. Or we would leave a_2's AGREE message without response, whereupon we would rely on the decision logic of the agent to resolve the problem (e.g. by having her assume failure after some time).

Our data-centric view affords us with additional ways of dealing with such problems. Firstly, because of our protocol semantics, the INFORM_INVALID option is easy to accommodate, as a_2 can still AGREE to any task contained in the INFORM_TASK message store. Secondly, the orchestrator could remove INFORM_TASK $(o, \{a_2, a_3\}, t_2)$ (as owner of this message) after receiving a REJECT from a_3, and a_2 would be able to anticipate that its previous AGREE message has become invalid (it could even be *del*eted by o if we used this type of call and gave the orchestrator appropriate permissions for this operation on messages not created by herself). Under these circumstances, not even the INFORM_INVALID message itself would be necessary, thus making the protocol even simpler. Finally, we could give a_3 permission to *add* INFORM_TASK messages (for example with possible alternative tasks that were not generated by o) or post *mod*ifications to t_2 in order to make the task acceptable for her, thus increasing the chances that successful agreement would be reached.

Thus, even though in principle the possible computations that can be jointly performed by agents are of course no different from the agent-centric view, our data-centric view allows much more flexibility in organising the interactions that lead to those computations, without requiring that the overall protocol needs to be significantly re-designed to accommodate additional functionality. For example, we could have orchestrators post arbitrary new tasks in an asynchronous way (for the same requests, or incrementally, as more potentially matching agents join), we could easily allow drivers to agree to several tasks in parallel, or let peers remove their previous requests if they are no longer interested in them.

5   Experimental Results

To establish whether the scalability and robustness we expect can actually be observed in a real-world implementation, we have developed a prototypical web-based system that runs the protocol depicted in Fig. 1, and evaluated it experimentally in the ridesharing domain. Our experiments below focus on the matchmaking and negotiation part of the protocol (from REQUEST to START), as this involves most dependencies among individual behaviours, and requires involves solving a complex combinatorial problem for the orchestrator agent o that involves calculating exponential numbers of possible rides presented to every driver and passenger. Instead of trying to get agreement or rejection to a single potential ride from every peer involved, our architecture enables us to constantly update all rides available to every peer in the system. We also use two further "non-standard" protocol operations: One is to automatically generate INFORM_INVALID messages for other participants when an agent REJECTS a ride, and the other is to delete all INFORM_TASK messages linked to a peer's request once a different ride for that peer has been agreed. Since in practice there is no global clock for synchronisation, all agents periodically poll the stores they are interested in (INFORM_TASK to check what the currently available rides are, and START/INFORM_INVALID to determine whether a ride has been agreed/can no longer be agreed). In terms of the execution engine, our implementation involves a single server which contains all message stores, and exposes operations on them through a simple RESTful Web API. The server runs Node.js, a non-blocking event-driven JavaScript library, and has separate processing queues associated with different message stores, which asynchronously process individual "platform jobs" for different client calls. Note that running the platform on a single server is not a requirement – in principle every message store could be located on a different server, including agent nodes that implement an HTTP interface.

Our first experiment examines the overall scalability of the platform. We create artificial "groups" of size k in a population of n agents such that all the requests inside a group match, and we can artificially control how many rides will be created. Our first experiment involves up to 10 groups of 6, 9, and 12 agents, i.e. a total of 60, 90, 120 agents, where the ratio of drivers d to passengers p is 1/2 (i.e. $d/p \in \{2/4, 3/6, 4/8\}$ for each group size). Note that the respective number of possible rides generated in each group is $(2^p - 1) * d$ as there is a different proposal for every subset of passengers, and the rides different drivers may offer to a group overlap. This means that 30/189/1020 rides have to be created for each group, i.e. the system has to deal with up to 10200 rides overall as we keep adding groups. Note also that, since all ride requests and agreements to rides occur in very close succession, the load of this system is similar to a real-world system that would experience this level of usage every few minutes (in reality, of course, users take much longer to check updates and respond), so it is in fact representative of a very large scale real-world application. Finally, to maximise the amount of messages exchanged and the duration of negotiation, drivers accept only the maximally sized ride, and passengers accept all rides. The top two plots in Fig. 3 show the average time taken in seconds (across all agents, and for 20 repetitions for each experiment, with error bars to indicate standard deviations) for matchmaking (REQUEST and INFORM_TASK) and negotiation (all further messages up to and including START), respectively. As can be seen from these plots, even though every agent has a built-in delay of 2 s between any two steps, even when there are 120 agents in the system, the average time it takes an agent to get information about all rides acceptable to her/complete the negotiation of a ride is around 50 s/80 s even in the largest configurations.

In the second experiment, we investigate the cumulative effect of adding delays and message failures on the total execution time of an entire negotiation for a ride, in order to assess how robust the system is. For this, we artificially increase the delay between any update an agent receives and its successive operation from 2 s to 5 s, 10 s, and 20 s. We use these artificial delays also to emulate failure, e.g. when network resources are temporarily unavailable. The bottom plot in Fig. 3 shows the results for this experiment, for a group size of 9 and 5 groups (45 agents in total), showing measurements for matchmaking, negotiation, and the total lifespan of an agent (from creation to agreement). As can be seen, the overall lifespan of an agent increases by a factor of 3 to 4 here when the delay increases by a factor of 10, which is a good indication that the system degrades gracefully under increasing perturbation. Moreover, what is interesting is that the time taken for negotiation, which involves the highest number of messages to the orchestrator (as all passengers accept all rides) only increases by a factor between 1.5 and 2. This is because the larger delays require less effort for matchmaking and computing rides, and the orchestrator has more time to process negotiation-related messages during these gaps. This nicely illustrates how separating the processing of different message stores leads to effective load balancing for any agent that has to engage in different interactions concurrently.

6   Related Work

The idea of coordinating distributed processes through a shared coordination medium is not new. It can be traced back at least to the blackboard systems [4] used in early distributed knowledge-based systems. In distributed computing, similar ideas led to coordination languages like LINDA [7]. While these systems initially involved either fixed sets of coordination primitives or built-in, application-dependent coordination strategies, they were later used in platforms like TuCSoN [11] to develop programmable behaviours for the coordination data structures. Our approach differs from this line of work in that we do not attempt to replace the protocol-based interaction models used in mainstream agents research. Instead, we maintain their advantages in terms of supporting complex specifications of sequential interactions and agent communication language semantics. Mapping these onto a data-driven architecture gives us the "best of both worlds", as it allows us to capture complex agent interactions while separating coordination from computation.

An architecture that takes a similar protocol-centric approach to the regulation of agent behaviours is OpenKnowledge [12], which allows declarative specifications of interaction protocols to be directly executed in an open, peer-to-peer platform. While its automation of executing protocols from their specification is more advanced here than in our approach, it involves agents effectively handing over control to coordinators that "run" the agent processes (the agent can still

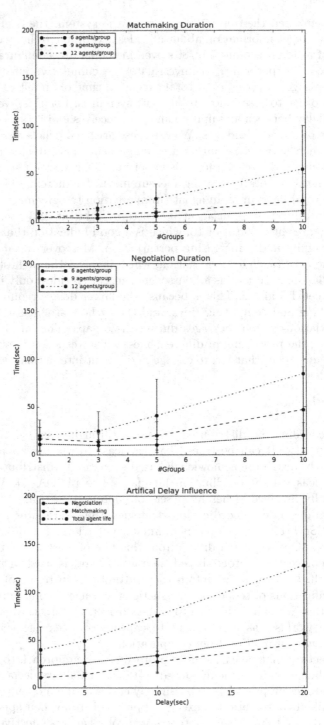

Fig. 3. Experimental results

make autonomous decisions regarding different choices available in the protocol, but the platform executes the protocol by invoking these local decision methods from outside). The difference to our approach is that we do not provide an execution platform that includes the agent processes themselves, but prefer to restrict the computational coordination process only to what is absolutely necessary.

Previous work most closely related to ours, however, is at the intersection of agents and service-oriented computing research. The authors of [2] present a method for mapping complex agent conversations to web services, without providing, however, a formal framework or a concrete implementation. Only very recently Singh [14,15] addressed the service-based view of agent protocols by proposing a formal language and a computational architecture that supports it. His approach bears close resemblance to our work: It considers protocols in terms of information schemas without any additional control flow constructs, defines semantics in terms of histories of past message exchanges, and proposes an architecture that enables agents to asynchronously and atomically process individual messages, supporting distributed interactions that have multiple loci of enactment. The main difference to our approach is that the semantics provided in this model does not take account of non-message actions and local state transitions, instead focusing more on protocol verification. Also, no quantitative performance results for an implementation of this system are presented.

It is worth mentioning, that, with the exception of [14], all of the above approaches involve some kind of *middleware* that relies on a bespoke communication architecture and platform that the agents must comply with. Moreover, while these platforms could be exposed, at least in principle, over normal Web APIs, agent designers would still have to be familiar with the specific languages used by them. In our framework, we do not only do away with such specific middleware. We also reduce the language specification for messages and constraints to a very general form, through constraints that are general variable restrictions, and messages with simple pre- and postconditions. As long as ontological agreement can be assumed regarding the semantics of individual variables and their domains (which is also a prerequisite for all of the above approaches), our APIs should be straightforward to use. In these respects, our work is heavily influenced by the REST paradigm [5], in that it uses ordinary Web resources as the means of exposing state to peers in order to coordinate the workflow between them. To our knowledge, there have been no attempts to formalise the semantics of this paradigm, and while our work does not aim to provide such semantics for the general case, it can be seen as a contribution toward a better overall understanding of REST itself.

7   Conclusions

In this paper, we have presented a data-driven architecture for coordinating agent collectives on the Web that is aimed at bridging the gap between work on agent interaction protocols and modern Web-based applications used commonly in areas such as social and collective computation. We presented a formal framework that allows us to specify the semantics of our architecture, and which allowed us to introduce functionality that is not available in normal agent-based systems platforms. Our experimental results with a prototypical implementation show that it can handle complex interactions in a lightweight way, producing minimal overhead while providing good scalability and robustness properties.

We summarise the main benefits of our approach: Firstly, there are no sequential distributed processes that need to rely on a standing line of communication, since all data operations are atomic, and can be easily repeated in case of failure. As our experiments show, the overhead of the additional link structure that has to be stored and the frequent "pull" operations from agents do not seem to affect performance significantly. Secondly, in a real Web deployment, we could directly benefit from the standard caching facilities of Web servers that can store frequently reused resources. Thirdly, coordination platforms can cross-check parallel interactions and apply global constraints to the overall interactions flexibly. Fourthly, since all operations are atomic, the decision logic can be devolved to components processing data in parallel whenever different steps are independent. This also provides guidance for designing agents' internal reasoning mechanisms, or for "splitting" functionality into several agents. Finally, message stores and the linkage between them provide a direct "data" view to the ongoing interactions at a global level, thus facilitating analysis, prediction, and ease of mapping to other structures such as provenance information. In fact, all operations in our architecture can easily be captured using standard formats like PROV[6], and our implementation supports this through full integration with a live PROV server.

As regards future work, on the practical side, we plan to focus on developing an automated procedure for generating implementations of our architecture directly from a given protocol specification. On the more theoretical side, we would like to develop formal procedures to detect and decouple different parts of a coordination protocol where these are not causally linked.

Acknowledgments. The research presented in this paper has been funded by the European Community's Seventh Framework Programme (FP7/2007–2013) under grant agreement n. 600854 *"SmartSociety – Hybrid and Diversity-Aware Collective Adaptive Systems: Where people meet machines to build smarter societies"* (http://www.smart-society-project.eu/).

References

1. Ahmad, S., Battle, A., Malkani, Z., Kamvar, S.D.: The Jabberwocky programming environment for structured social computing. In: Proceedings of the 24th Annual ACM Symposium on User Interface Software and Technology (UIST 2011), pp. 53–64, Santa Barbara, CA (2011)
2. Ardissono, L., Goy, A., Petrone, G.: Enabling conversations with Web services. In: Rosenschein, J.S., Sandholm, T., Wooldridge, M., Yokoo, M. (eds.) Proceedings of the Second International Joint Conference on Autonomous Agents and Multiagent Systems (AAMAS 2003), pp. 819–826 (2003)

---

[6] See http://www.w3.org/TR/prov-overview/.

3. Chopra, A.K., Artikis, A., Bentahar, J., Colombetti, M., Dignum, F., Fornara, N., Jones, A.J.I., Singh, M.P., Yolum, P.: Research directions in agent communication. ACM Trans. Intell. Syst. Technol. **4**(2), 1–23 (2013)

4. Englemore, R., Morgan, T. (eds.): Blackboard Systems. Addison-Wesley, Reading (1988)

5. Fielding, R.T., Taylor, R.N.: Principled design of the modern web architecture. ACM Trans. Internet Technol. **2**(2), 115–150 (2002)

6. Fikes, R.E., Nilsson, N.J.: STRIPS: a new approach to the application of theorem proving to problem solving. Artif. Intell. **2**(3–4), 189–208 (1971)

7. Gelernter, D., Carriero, N.: Coordination languages and their significance. Commun. ACM **35**(2), 97–107 (1992)

8. Law, E., Von Ahn, L.: Human Computation. Synthesis Lectures on Artificial Intelligence and Machine Learning. Morgan & Claypool Publishers, San Rafael (2011)

9. Little, G., Chilton, L.B., Goldman, M., Miller, R.C.: TurKit: human computation algorithms on mechanical turk. In: Proceedings of the 23nd Annual ACM Symposium on User Interface Software and Technology (UIST 2010), pp. 57–66 (2010)

10. Malone, T.W., Laubacher, R., Dellarocas, C.: The collective intelligence genome. Sloan Manag. Rev. **51**(3), 21–31 (2010)

11. Omicini, A., Zambonelli, F.: Tuple centres for the coordination of internet agents. In: Proceedings of the 4th ACM Symposium on Applied Computing, San Antonio, TX, February 1999

12. Robertson, D., et al.: Models of interaction as a grounding for peer to peer knowledge sharing. In: Dillon, T.S., Chang, E.J., Meersman, R., Sycara, K. (eds.) Advances in Web Semantics I. LNCS, pp. 81–129. Springer, Heidelberg (2007)

13. Rovatsos, M.: Multiagent systems for social computation. In: Lomuscio, A., Scerri, P., Bazzan, A., Huhns, M. (eds.) Proceedings of the 13th International Joint Conference on Autonomous Agents and Multiagent Systems (AAMAS 2014), Paris, France, 5th–9th May 2014

14. Singh, M.P.: LoST: local state transfer - an architectural style for the distributed enactment of business protocols. In: Proceedings of the 9th International Conference on Web Services (ICWS), pp. 57–64, Washington, DC (2011)

15. Singh, M.P.: Semantics and Verification of Information-Based Protocols. In: Conitzer, V., Winikoff, M., Padgham, L., Van der Hoek, V. (eds.) Proceedings of the Eleventh International Conference on Autonomous Agents Multiagent Systems (AAMAS 2012), pp. 1149–1156, Valencia, Spain, June 4–8, 2012

Negotiating Privacy Constraints in Online Social Networks

Yavuz Mester, Nadin Kökciyan$^{(\boxtimes)}$, and Pınar Yolum

Department of Computer Engineering, Bogazici University,
34342 Bebek, Istanbul, Turkey
{yavuz.mester,nadin.kokciyan,pinar.yolum}@boun.edu.tr

Abstract. Privacy is a major concern of Web systems. Traditional Web systems employ static privacy agreements to notify its users of how their information will be used. Recent social networks allow users to specify some privacy concerns, thus providing a partially personalized privacy setting. However, still privacy violations are taking place because of different privacy concerns, based on context, audience, or content that cannot be enumerated by a user up front. Accordingly, we propose that privacy should be handled per post and on demand among all that might be affected. To realize this, we envision a multiagent system where each user in a social network is represented by an agent. When a user engages in an activity that could jeopardize a user's privacy (e.g., publishing a picture), agents of the users negotiate on the privacy concerns that will govern the content. We employ a negotiation protocol and use it to settle differences in privacy expectations. We develop a novel agent that represents its user's preferences semantically and reason on privacy concerns effectively. Execution of our agent on privacy scenarios from the literature show that our approach can handle and resolve realistic privacy violations before they occur.

1 Introduction

Privacy has long been accepted as an important concept in developing and running software. A typical software publishes its privacy agreement, which a user accepts. A similar pattern applies to online social networks, with additional settings for customizing the policy. For example, a user can choose to share content only with her friends, whereas another user may choose to share content with everyone in the system.

Users' privacy requirements or expectations are often conflicting. While this is not a problem in regular software in which transactions are independent from each other, it creates concerns in social networks where users can manipulate content by tagging other users, resharing, and so on. Thus, content could become available to a broad audience without the consent of some of the people involved. Consider a user who posts a picture of herself with one of her friends. The posting, apparently harmless to the former, may be considered inappropriate by the latter [1].

This work has been supported by TUBITAK under grant 113E543.

F. Koch et al. (Eds.): CARE-MFSC 2015, CCIS 541, pp. 112–129, 2015.
DOI: 10.1007/978-3-319-24804-2_8

Facebook, for instance, deals with such issue by allowing users to report pictures as inappropriate and by informing a user that the tagged person is unhappy about the content and would prefer to see it removed. However, the informed user is free to decide what to do about it. This process of reporting inappropriate content and removing it requires human interaction, takes time, and often does not prevent the content from being exposed to a large audience before it is removed [7]. This example illustrates various desirable properties of privacy protection:

Automation: Privacy protection calls for automated methods. Considering the volume of social media transactions that are done every hour, it is not plausible for humans to discuss every content that is related to them in person. Hence, an agent that represents a user is needed to keep track of its user's preferences and policies and act on behalf of them, accordingly.

Fairness: If a particular post is deemed private for a user, one approach is to take it down from the system completely. This all-or-nothing approach is simple but leaves the party who wants to keep the post up in a disadvantaged position. Instead, it is best if the users identify what is wrong with the post (in terms of privacy) and improve those aspects only. For example, if the text of a picture post is causing a privacy violation, the text can be removed while keeping the picture.

Concealment of Privacy Concerns: Privacy preferences may reveal personal information that should be kept private from others. That is, a user might say that she does not like the text of a post, but she does not need to identify why that is the case. Ideally, a proposed approach to privacy protection should keep users' privacy concerns private, without revealing users' privacy concerns as a whole.

Protection Before Exposure: Contrary to existing systems, such as Facebook, where a privacy violation is caught after it happens, the fact that a content is private should be identified before the content is put up online. Otherwise, there is a risk that the content reaches unintended audience. Hence, a proposed approach should be in place before the content is put up rather than after.

Accordingly, this paper proposes an agreement platform for privacy protection that addresses the above properties. One recent study has shown that many times invading a friend's privacy on a social network is accidental and the user actually invading the privacy is not aware of it. When the user is notified for the situation, many of them choose to put down a content rather than breaching their friend's privacy [12]. This naturally suggests that if the users could have agreed on the content before it went up, then many privacy leakages could have been avoided to begin with. We exploit this idea by developing a platform where the agents interact to reach a consensus on a post to be published. We assume that each user is helped by a software agent to manage her privacy. The agent is aware of the user's privacy concerns and expectations but also knows about the user's social network, such as her friends. When a user is about to post a new content, she delegates the task to her agent. The agent reasons on behalf of the user to decide which other users would be affected by the post and contacts those users' agents. The negotiation protocol we develop enables agents to discuss their users' constraints and agree on a suitable way to post the content such that none of the users' privacy is violated. We show the applicability of our approach on example scenarios from the literature.

The rest of this paper is organized as follows: Sect. 2 develops our negotiation framework and shows how the agents use it to reach agreements. Section 3 describes our agent architecture with an emphasis on semantic representation and reasoning that it can do for its user. Section 4 evaluates our approach first by using different scenarios from the literature and then comparing it to prominent approaches from the literature based on the defined criteria above. Finally, Sect. 5 discusses our work in relation to recent work on negotiation and privacy.

2   Negotiation Framework

We propose a negotiation framework for privacy, PRINEGO, where users are represented by agents. An agent is responsible for keeping track of its user's privacy constraints and managing its user's privacy dealings with others. Before sharing a post, an agent decides if a post could violate other users' privacy (e.g., those involved in a picture). For this, an agent interacts with other agents to negotiate on a mutually acceptable post. When an agent receives a post request from a negotiator agent, it evaluates if the content would be acceptable for its user in terms of the media it contains, post audience, and so on. If the post request is not acceptable, that agent returns a rejection reason (e.g., audience should not contain a certain individual) to help the negotiator agent in revising its initial request. Once the negotiator agent collects all the rejection reasons from other agents, it revises its post request accordingly (e.g., removing the mentioned individual from the audience). A post negotiation terminates when all agents agree on a content or it reaches the maximum number of iterations set by the negotiator agent.

2.1   A Negotiation Protocol

Agents negotiate by employing a negotiation protocol. Our proposed negotiation protocol is in principle similar to existing negotiation protocols, which are used in e-commerce [6]. However, there are two major differences. First, contrary to negotiations in e-commerce, the utility of each offer is difficult to judge. For example, in e-commerce, a seller could expect a buyer to put a counter-offer with a lower price than that it actually made in the first place. Here, however it is not easy to compare two offers based on how private they are. Second, partly as a result of this, counter-offers are not formulated by the other agents but instead the negotiator agent collects the rejection reasons from other agents to update its initial offer.

There are two important components in our negotiation protocol: a *post request* and a *response*. A *post request* is essentially a post, which can include media, text, location, tagged people, audience and so on. Since this post has not been finalized yet (i.e., put up online), we consider this as being requested.

Hence, agents try to negotiate on a post request. A post request must contain information about: (1) owner of the post request, (2) a post content (text, medium and so on) to be published, and (3) a target audience. A *response* is generated when an agent receives a post request. A response must contain: (1) owner of the response, (2) a response code ("Y" for accept, "N" for reject), and (3) a rejection reason, which is optional. We explain rejection reason types in Sect. 2.3. We use Example 1 that is inspired from Wishart *et al.* [14] as our running example.

Example 1. Alice would like to share a party picture with her friends. In this picture, Bob and Carol appear as well. However, both of them have some privacy concerns. Bob does not want to show party pictures to his family as he thinks that they are embarrassing. Carol does not want Filipo to see this picture because she did not tell him that she is a bartender.

In Example 1, Alice instructs her agent to share a post by specifying her party picture where she tags Bob and Carol, and sets the audience to her friends. Alice's agent may decide to negotiate the post content with Bob and Carol as they are tagged in the picture. Then Alice's agent should initialize a *post request* (i.e., offer in our negotiation protocol) and send it to agents of Bob and Carol. These agents should then evaluate that post request with respect to their respective owner's privacy concerns, and create a *response*. Their responses should make it clear whether Bob and Carol accept or reject that post request, and ideally specify a reason in case of a rejection.

2.2 NEGOTIATE Algorithm

Algorithm 1 presents our algorithm for negotiating privacy constraints. The algorithm tries to finalize a post request iteratively, while negotiating with other agents. Basically, at each iteration, the negotiator agent decides which agents to negotiate with, collects their responses after that they evaluate the corresponding post request, and revises the initial post request if any rejection response is received and the negotiation continues with the revised post request. Whenever all agents agree, our algorithm returns a post request to be shared on the social network. If an agreement cannot be reached after a predefined maximum number of iterations, then the negotiator agent returns the latest post request without negotiating it further.

Our algorithm takes four parameters as input: (1) the newly created post request (p), (2) the media selected by the owner as alternatives to the original medium ($altM$), (3) the current iteration index (c), (4) the maximum number of negotiation iterations (m). When the algorithm is first invoked, p should contain the owner's original post request. If the owner specifies any alternative media, $altM$ should contain them. c should be 1, and m can be configured by the owner. Our algorithm returns the final post request resulting from the negotiation. We use *responses* to keep the incoming responses of other agents. If *responses* contain a rejection, the negotiator agent inspects the reasons included

---

Algorithm 1. p NEGOTIATE($p, altM, c, m$)

Require: p, the post request to be negotiated
Require: $altM$, the owner's choice of alternative media
Require: c, current iteration index
Require: m, maximum number of iterations
Return: the final post request resulting from the negotiation
1: **if** $c > m$ **then**
2:     **return** p
3: **else**
4:     $responses \leftarrow \emptyset$
5:     **for all** $agent \in$ DECIDEAGENTSTONEGOTIATE(p) **do**
6:         $response \leftarrow agent.$ASK($p$)
7:         $responses \leftarrow responses \cup \{response\}$
8:     **end for**
9:     **if** $\forall r \in responses, r.responseCode =$ "Y" **then**
10:         **return** p
11:     **else**
12:         $p' \leftarrow$ REVISE($p, altM, c, m, responses$)
13:         **if** $p' \neq NULL$ **then**
14:             **return** NEGOTIATE($p', altM, c + 1, m$)
15:         **else**
16:             **return** $NULL$
17:         **end if**
18:     **end if**
19: **end if**

---

in *responses* to revise its initial post request (p')[1]. There are three auxiliary functions used in our algorithm, each of which can be realized differently by agents:

- DECIDEAGENTSTONEGOTIATE takes p and decides which agents to negotiate with. A privacy-conscious agent may decide to negotiate with every agent mentioned or included in the post text and medium components of a post request[2], whereas a more relaxed agent may skip some of them. We provide more details of DECIDEAGENTSTONEGOTIATE function in Sect. 3.2.
- ASK is used to ask agents to either accept or reject p. Each agent evaluates p according to their owners' privacy concerns. We explain how our semantic agent evaluates a post request in Sect. 3.3.
- REVISE is used to revise a post request with respect to *responses* variable that includes rejection reasons. This function may also return $NULL$ at any iteration, which indicates a disagreement. The details of REVISE function are provided in Sect. 3.4.

---

[1] p' is similar to "counter-offer" in other negotiation protocols with the difference that the negotiator agent makes it.

[2] We assume that mentioned or included people information can be inferred from post text and medium by employing textual and facial recognition algorithms.

NEGOTIATE is a recursive algorithm, which starts by checking whether the current iteration index (c) exceeds the allowed maximum number of iterations (m) (line 1). If this is the case, then the algorithm returns the current p without further evaluation (line 2). Otherwise, the algorithm negotiates p as follows. *responses* variable is set to an empty set (line 4). In order to start the negotiation, the algorithm first computes which agents to negotiate with (line 5). Then, the algorithm asks each such agent to evaluate p (line 6), and adds each incoming response to *responses* (line 7). After that, the algorithm checks whether there is an agreement (line 9). This is simply done by inspecting the response codes; a response code "Y" means an acceptance. If all agents accept p, then the algorithm returns p (line 10). Otherwise, the algorithm calls the auxiliary function REVISE to generate the revised post request p' (line 12). If p' is not $NULL$ (line 13), then p' has to be negotiated. Thus, the algorithm makes a recursive call by providing p' as the first argument (line 14), and c is incremented by 1 to specify the next iteration number (line 14). Otherwise (line 15), the algorithm returns $NULL$ (line 16), which indicates that no agreement has been reached.

2.3 Rejection Reasons

Agents may reject any post request and optionally provide the underneath reason along with the response. A rejection reason should specify the field of discomfort, such as audience, post text, and properties of the medium. Providing a reason for rejection is important because it gives the negotiator agent the opportunity to revise its request to respect the other agents' privacy constraints. Generally, an agent may have multiple reasons to reject a post request. We design our protocol to accept the reasons one by one to ease the processing. For example, a user may not want to reveal her party pictures to her colleagues. Whenever her agent is asked about a post request including such a medium and target audience, the agent should choose reasons related to medium or audience, but not both. This is a design decision to simplify the revision process. The post request is refined iteratively based on the rejection reasons collected in each iteration. Getting one reason from each agent may already result in a dramatic change in the issued post request (e.g. the medium can be altered and the new one may be tagged with a different set of people, such that different agents has to be consulted in the next iteration). The other rejection reasons may not be valid after the change, thus no need to be evaluated up front. If they are still valid, rejections may rise again in the next iterations and be evaluated. By adopting such a restriction, the privacy rules themselves have some degree of privacy as well. Only an implication of a privacy concern is exposed at a time. We provide more details about reason selection in Sect. 3.3.

3 A Semantic PRINEGO Agent

Our negotiation framework, PRINEGO, is open in the sense that it enables agents that are built by different vendors to operate easily. PRINEGO mainly requires agents to implement the negotiation protocol given in Algorithm 1 and conform with an agent skeleton. The agent skeleton is meant to describe the minimal set of functionalities that should be handled by an agent in order to qualify. The agents should be able to have the following functionalities: (1) Each agent should represent knowledge about the social network, in terms of relations among users, context of posts, user preferences, and so on. (2) Each agent should be able to implement the three methods of Algorithm 1; that is, DECIDEAGENTSTONEGOTIATE, ASK, and REVISE. Hence, an agent can start a post negotiation. (3) Each agent should be able to implement the EVALUATE algorithm so that it can evaluate a post request and decide if it is acceptable. The following sections explain how these are handled in our semantic PRINEGO agent.

3.1   A Social Network Ontology

We develop PRINEGO ontology to represent the social network as well as the privacy constraints of users. Each user has her own ontology that keeps information about her relationships, content and privacy concerns. An agent has access to its user's ontology and our negotiation framework uses ontologies to model interactions between agents.

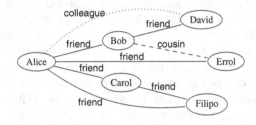

Fig. 1. Relation properties in PRINEGO ontology

The Social Network Domain. A social network[3] consists of *users* who interact with each other by sharing some *content*. Each user is connected to another user via *relations* that are initiated and terminated by users. The social network domain is represented as an ontology using Web Ontology Language (OWL). In our negotiation framework, each `Agent` interacts with other agents by sending post requests. A `PostRequest` consists of a post that is intended to be seen by a target `Audience`, *hasAudience* is used to relate these two entities. An audience consists of its agents and we use *hasAudienceMember* to specify members of an audience. A `PostRequest` can contain some textual information `PostText` that may mention people or some `Location` (e.g., `Bar`). For this, we use *mentionsPerson* and *mentionsLocation* respectively. Moreover, a post request

---

[3] We denote a `Concept` with text in mono-spaced format, a *relationship* with italic text, and an `:instance` with a colon followed by text in mono-spaced format.

can include some `Medium` (e.g., `Picture` and `Video`). *hasMedium* is a property connecting `PostRequest` and `Medium` entities. A medium can give information about people who are in that medium, or location where that medium was taken. Such information is described by *includesPerson* and *includesLocation* properties respectively. Moreover, *isDisliked* is a boolean property that a user can use to dislike a medium.

Relation Properties. In a social network, users are connected to each other via various relationships. Each user labels his social network using a set of relationships. We use *isConnectedTo* to describe relations between agents. This property only states that an agent is connected to another one. The sub-properties of *isConnectedTo* are defined to specify relations in a fine-grained way. *isColleagueOf*, *isFriendOf* and *isPartOfFamilyOf* are properties connecting agents to each other in the social network of a user. They are used to specify agents who are colleagues, friends and family respectively. Figure 1 depicts the social network for our motivating and evaluation examples. Here, a node represents an agent and the edges are the relations between agents. Each edge is labeled with a relation type. For simplicity, we use friend, cousin and colleague keywords in the figure. All edges are undirected as the relations between agents are symmetric. For example, Alice and Carol are friends of each other.

Context. Context is an essential concept in understanding privacy. Many times, context is simply interpreted as location. The idea being that your context can be inferred from your location; e.g., if you are in a bar, then your context can be summarized as leisurely. Sometimes, the location is combined with time to approximate the context better. However, even location and time combined does not suffice to understand context [9]. Contrast a businessman going to a bar at night and the bartender being there at the same time. Time and location-wise their situation is identical, but most probably their contexts are different as the first one might be there to relax; the second to work.

This idea of context applies similarly on the content that is shared. A context of a picture might be radically different for two people who are in the picture. Hence, it is not enough to associate a context based on location only. Further information about a particular person needs to be factored in to come up with a context. Our ontology contains various `Contexts` that can be associated with a post request for a given person. Hence, each agent infers context information according to the context semantics associated with its user. Following the above example, a medium taken in a bar may reveal `EatAndDrink` context for the businessman and `Work` context for the bartender. We use *isInContext* to associate context information to a medium.

Protocol Properties. Each post request is owned by an agent that sends a post request to other agents. In a post negotiation, an agent may accept or reject a post request according to its user's privacy concerns. In the case where an agent *rejects* a post request, *rejectedIn* defines which concept (Medium, Audience or Post text) causes the rejection. The user can provide more detailed information about the rejection reasons by the use of *rejectedBecauseOf* and *rejectedBecauseOfDate* properties. In OWL, there is a distinction between object and data properties. Object properties connect pairs of concept instances while data properties connect instances with literals. Hence, *rejectedBecauseOf* is used if a post is rejected because of a concept (e.g., `Audience`) while we use *rejectedBecauseOfDate* to refer to a date literal (e.g., 2014-05-01). For example, a user can reject a post request because of an audience which includes undesired people or a medium where the user did not like herself. We explain protocol properties in detail in Sect. 3.3.

3.2   DecideAgentsToNegotiate Algorithm

In many negotiation problems, an agent already knows with whom to negotiate. However, in the social network domain, an agent can decide with whom to negotiate. Consider a picture of Alice and Bob taken in Charlie's office. When Alice's agent is considering of putting this up, it might find it useful to negotiate with Bob only but not with Charlie. A different agent might have found it necessary to also negotiate with Charlie; even though he is not in the picture himself, his personal information (e.g., workplace) will be revealed.

The choice to pick agents to negotiate with can also be context-dependent. In certain contexts, an agent can prefer to be more picky about privacy and attempt to negotiate with all that are part of a post. For example, if the post contains information that could reveal medical conditions about certain people, an agent might be more careful in getting permissions from others beforehand. In our case, our agent chooses to negotiate with everyone that is tagged in a post.

3.3   Evaluate Algorithm

Once an agent receives a post request, it evaluates whether it complies with its user's privacy concerns. A post request includes various types of information: media, post text, target audience, and so on. Essentially, any of these points could be unacceptable for the agent that receives the request, which provides information about rejected attributes of the post request in case of a rejection. This can be considered as an explanation for the negotiator agent who revises its post request as accurately as possible so that it may be accepted by other agents in the following iterations.

Privacy Rules. The privacy rules reflect the privacy concerns of a user. In a privacy rule, the user declares what type of post requests should be rejected at negotiation time. A privacy rule may depend on a specific location, context, relationship or any combination of these. Each agent is aware of the privacy concerns of its user. At negotiation time, the privacy rules are processed by an agent in order to decide how to respond to incoming post requests. For this, we augment PriNego with the privacy rules by the use of Semantic Web Rule

Language (SWRL) rules for more expressiveness [5]. Description Logic reasoners can reason on ontologies augmented with SWRL rules. Each rule consists of a *Head* and a *Body* and is of the form *Body* $\implies$ *Head*. Both the body and head consist of positive predicates. If all the predicates in the body are true, then the predicates in the head are also true. We use SWRL rules to specify Privacy Rules (P) in our negotiation framework. In a privacy rule, concepts and properties that are already defined in PRINEGO are used as predicates in rules. Moreover, we use protocol properties in the head of a rule. Here, we consider privacy rules to specify why a particular post request would be rejected. Any post request that is not rejected according to the privacy rules is automatically accepted by an agent. An agent can reject a post request without providing any reason. In this case, *rejects* is the only predicate observed in the head of a privacy rule. On the other hand, if an agent would like to give reasons about a rejection, then *rejectedIn* property is used to declare whether the rejection is caused by a medium, a post text or an audience. Furthermore, *rejectedBecauseOf* is used to specify more details about the rejection. For example, an audience can be rejected in a post request because of an undesired person in the audience. Or a medium can be rejected in a post request because of the location where the medium was taken. Here, we assume that privacy concerns of a user are already represented as privacy rules in her ontology.

In Example 1, Bob and Carol have privacy constraints that are shown in Table 1. P_{B_1} is one of Bob's privacy rules. It states that if a post request consists of a medium in Party context and has an audience member from Bob's family then Bob's agent rejects the post request because of the audience and this audience member becomes a rejected member in the corresponding post request. Carol's privacy rule (P_{C_1}) states that if a post request consists of a medium in Work context and if Filipo is an audience member in the audience of the post request then Carol's agent rejects the post request because of two reasons. Filipo is an undesired person in the audience hence her agent rejects the post request because of the audience and Filipo becomes a rejected member in the post request. The second reason is that Carol does not want to reveal information about her work, her agent rejects the post request because of the context as the medium discloses information about Work context.

As shown in P_{C_1}, multiple components of a post request may be marked as rejected components in the head of a rule; e.g., medium and audience. Moreover, each component may be rejected because of one or more reasons. In P_{C_1}, each component is rejected because of one reason; e.g., context and audience member respectively. When multiple components of a post request are marked as rejected as the result of ontological reasoning, the agent has to decide on the component to share as a rejection reason. In this work, we consider reasons that would require minimal change in the initial post request. For this, we adopt a hierarchy between the components as follows: (i) audience, (ii) post text and (iii) medium. However, this behavior can vary from agent to agent. When P_{C_1} fires, the post request will be rejected by the reasoner because of the medium and the audience. Then, regarding our hierarchy, :carol first checks whether there is any rejection caused by the audience and she finds one. Hence, it rejects the audience because of :filipo. In another words, :carol will share this reason (and not the context reason) with the negotiator agent.

Table 1. Privacy rules (P) as SWRL rules

P_{A_1}:	*hasAudience*(?postRequest, ?audience), *hasAudienceMember*(?audience, ?audienceMember), Leisure(?context), *hasMedium*(?postRequest, ?medium), *isInContext*(?medium, ?context), *isColleagueOf*(?audienceMember, :alice) $\implies$ *rejects*(:alice, ?postRequest), *rejectedIn*(?audience, ?postRequest), *rejectedBecauseOf*(?audience, ?audienceMember)
P_{B_1}:	*hasAudience*(?postRequest, ?audience), *hasAudienceMember*(?audience, ?audienceMember), Party(?context), *hasMedium*(?postRequest, ?medium), *isInContext*(?medium, ?context), *isPartOfFamilyOf*(:bob, ?audienceMember) $\implies$ *rejects*(:bob, ?postRequest), *rejectedIn*(?audience, ?postRequest), *rejectedBecauseOf*(?audience, ?audienceMember)
P_{B_2}:	*hasMedium*(?postRequest, ?medium), *isDisliked*(?medium, true) $\implies$ *rejects*(:bob, ?postRequest), *rejectedIn*(?medium, ?postRequest), *rejectedBecauseOf*(?medium, :bob)
P_{C_1}:	*hasAudience*(?postRequest, ?audience), *hasAudienceMember*(?audience, :filipo) *hasMedium*(?postRequest, ?medium), Work(?context), *isInContext*(?medium, ?context) $\implies$ *rejects*(:carol, ?postRequest) *rejectedIn*(?medium, ?postRequest), *rejectedBecauseOf*(?medium, ?context), *rejectedIn*(?audience, ?postRequest), *rejectedBecauseOf*(?audience, :filipo)
P_{C_2}:	*hasMedium*(?postRequest, ?medium), *dateTime*(?t), *equal*(?t, "2014-05-01T00:00:00Z"), *hasDateTaken*(?medium, ?t) $\implies$ *rejects*(:carol, ?postRequest), rejectedIn(?medium, ?postRequest), *rejectedBecauseOfDate*(?medium, ?t)

3.4   Revise Algorithm

In case a post request is rejected by one or more agents, the negotiator agent needs to revise the post request with respect to the responses received. However, each agent is free to decide if it wants to credit a rejection reason or to ignore it. This is correlated with the fact that some people may have more respect to others' privacy, whereas some may be more reluctant to it. Moreover, the way the agents honor a rejection reason can vary as well. For example, consider a case where an agent rejects a post request because its owner is not pleased with his appearance in the picture. The negotiator agent can either alter the picture or remove it completely. In case it is altered, the new picture may or may not include the previously rejecting owner. The agents have the freedom of revising as they see appropriate.

An important thing to note is that the rejection reasons cannot possibly conflict with each other. This is guaranteed by two design decisions included in PriNego. First, the privacy concerns define only the cases where an agent should reject a post request. Second, the agents cannot reject a post request because it does not have some desired attributes (e.g., the audience should have also included some other person). In such a case, the agent may initiate another negotiation with a new post request that puts that person into the audience; i.e., resharing through negotiation is possible at any time. In this manner, a negotiator agent does not need to worry about conflicts and can handle rejection reasons in a suitable manner. Our semantic agent honors every rejection reason since we know that it will not create any conflict with other reasons. Our algorithm discards undesired audience members if any from the audience, and removes the text content if rejected by some agent. In Sect. 4.1, we show how our approach works step by step on Example 1.

When the medium component is rejected, our algorithm inspects all of the rejection reasons gathered during the negotiation, and clusters them under these five groupings: undesired included people, locations, and media dates, self-disliked mediums, context-disliked mediums. Then, our algorithm inspects each medium in the alternative media selection of the owner to find a suitable medium. The suitable medium should comply with other agents' privacy concerns. The approach we use to select a medium comes in handy for some cases and one such example is carried out in Sect. 4.2. After a revision, our algorithm checks whether the resulting post request is still reasonable. A post request is not reasonable if it does not have any audience or a content (neither text nor medium content). Thus, it is possible to result in a disagreement not because the rejection reasons may conflict with each other, but because the sanity of the post request may be lost after the revision process.

Our algorithm can be extended so that all the iterations for all the negotiations can be taken into consideration. Machine learning methods may suit here to estimate the possibility of a post request to be rejected with the help of past experience. Such enhancements may lead to intelligent revisions, and thus increase the chance and the speed of converging to an agreement dramatically. For example, a negotiator agent can learn the behavior of another agent (e.g., an agent that accepts every post request) and then decides not to ask that agent as it already knows that it will accept.

4 Evaluation

We implement the semantic PRINEGO agent that was detailed in Sect. 3 using Java and the Spring framework. We simulate the communication between the semantic agents with RESTful web services such that each agent has exactly two web services: one to be asked by the owner to negotiate a desired post, and the other to be asked by other agents to evaluate a post request. We use the OWL API [4] to work with ontologies and Pellet as the reasoner [10]. Each semantic agent uses an ontology as described in Sect. 3.1. The ontology is used as a knowledge base and also contributes into the reasoning of privacy rules (Sect. 3.3). When asked to evaluate a post request via a web service call, the agent first puts the incoming post request into its ontology and then makes ontological reasoning on the post request against the owner's privacy constraints.

The agent can run from a Web browser and additionally it can run as an Android application that we have developed[4]. Our mobile application is integrated with a real social network, Facebook. A user logs into our application by

---

[4] PriNego Mobile can be accessed at http://goo.gl/kF9svI.

providing her Facebook credentials. To initialize a post request, she selects a picture from her device and tags some of her Facebook friends. Moreover, she sets an audience for her post request. Our application starts a negotiation between the negotiator agent and the agents of tagged users. Once the negotiation is done, the negotiator agent shares the resulting post.

We evaluate our proposed approach using three scenarios from the literature. For this, we create six test users on Facebook, each of which has one of the PRINEGO agents we created, namely :alice, :bob, :carol, :david, :errol, and :filipo. We log into our mobile application with the appropriate test user and share a post request as suggested by each scenario. For each post request, PRINEGO is used for negotiating with other agents and the negotiator agent shares the agreed upon post requests on Facebook. Then, we examine the results to see how the agents reach agreements. We provide a walk-through examination of PRINEGO on Example 1 in Sect. 4.1 and then discuss two more scenarios in Sect. 4.2.

4.1 A Walk-Through of PRINEGO

In Example 1, Alice wants to share a post request p, which includes a party picture. Bob and Carol are tagged in this picture. The audience of the post request is set to Bob, Carol, Errol and Filipo. In order to publish the post request, Alice's agent (:alice) would like to negotiate with other agents. :alice does not have any other alternative media to share and is allowed to finalize a negotiation in five iterations as configured by Alice. :alice invokes NEGOTIATE $(p,[],1,5)$. It decides to negotiate with all the tagged agents in the post request, namely :bob and :carol, as a result of DECIDEAGENTSTONEGOTIATE (p). Then, :alice asks both of them by invoking ASK (p). :bob checks whether p is compatible with its privacy rules P_{B_1} and P_{B_2}. It notices that p includes a picture taken in **Party** context and Errol, who is part of Bob's family, is included in the audience of p. This condition fires one of the rules, P_{B_1}. As a consequence, :bob rejects p, the audience is rejected in p because of Errol, who is in the audience. :alice keeps :bob's rejection reason {-Errol}. On the other hand, :carol has two privacy rules P_{C_1} and P_{C_2}. :carol computes that p violates Carol's privacy rule P_{C_1}. p includes a picture, which is in **Work** context for Carol, moreover Filipo is in the audience of p hence :carol rejects p because of two reasons. It rejects p because of: (i) the audience, which includes Filipo, (ii) the **Work** context. :carol prioritizes these reasons (see Sect. 3.3) and chooses {-Filipo} as the rejection reason. The set of rejected reasons then becomes {-Errol, -Filipo}, and :alice wants to make a revision on p. :alice invokes REVISE $(p,[],1,5,\{-Errol, -Filipo\})$, which in turn revises p by removing the undesired people from the audience. The audience of p' (the revised post request) is set to Bob and Carol. :alice invokes NEGOTIATE $(p',[],2,5)$ to negotiate p'. :alice asks :bob and :carol, and finally both of them accept p', because neither of their rules are dictating otherwise. Thus, :alice finalizes the negotiation in two iterations and the resulting post request is harmful for neither Bob's nor Carol's privacy.

Table 2. Iteration steps for Example 2 where :bob starts the negotiation

Iter	Content	Audience	Asked Agents	Evaluate	Response
1	beach picture of Alice	A, D, E	:alice	:alice → P_{A_1}	:alice → -David
2	beach picture of Alice	A, E	:alice	:alice → N/A	:alice → ✓

Table 3. Iteration steps for Example 3 where :alice starts the negotiation

Iter.	Content	Audience	Asked Agents	Evaluate	Response
1	pic_1	B, C, E, F	:carol	:carol → P_{C_2}	:carol → -date
2	pic_2	B, C, E, F	:carol, :bob	:carol → N/A, :bob → P_{B_2}	:carol → ✓, :bob → -self
3	pic_3	B, C, E, F	:carol, :bob	:carol → N/A, :bob → N/A	:carol → ✓, :bob → ✓

4.2 PRINEGO in Action

In this section, we show how our semantic agents negotiate with each other in Examples 2 and 3. Tables 2 and 3 summarize the negotiation steps. We use initial letters for users in the audience description (e.g., A represents Alice). Example 2 is inspired from the work of Squicciarini *et al.* [11]. In this example, a user's privacy is compromised because of a friend sharing some content about this user.

Example 2. Alice and David are friends of Bob. Bob organizes a party where Alice and David meet each other. David offers Alice a job in his company and she accepts. One day Bob shares a beach picture of Alice with his connections. David could view this picture. Alice is worried about jeopardizing her position in David's company and asks Bob to remove her beach picture.

In Example 2, :bob starts the negotiation as Bob wants to share a beach picture of Alice. In the first iteration, :bob asks :alice, which rejects this post request according to P_{A_1} with the reason that the audience includes David. Then, :bob revises the post request by removing David from the audience (the content remains untouched) and sends the updated post request to :alice again. This time :alice accepts, and :bob terminates the negotiation as an agreement is reached.

Example 3. Alice would like to share a picture, which was taken on May 1, 2014 with her friends. Carol is tagged in this picture and she does not want to show any picture that was taken on May 1, 2014. Alice decides on sharing another picture where Carol and Bob are both tagged. This time Bob does not like himself in this picture. Alice finds another picture of Carol and Bob and finally they all agree to share.

Differently in Example 3, the rejection reasons are about the medium content of the post request, and the negotiator agent makes use of the alternative media.

In the first iteration, :carol rejects the picture because of its date taken. Then, :alice alters the picture and the new picture includes Carol as well as Bob. Thus this time :alice chooses to ask not only :carol but also :bob. However, :bob rejects since Bob marks the picture as self-disliked, although the new picture does not violate :carol's privacy constraints. :alice asks another picture of Bob and Carol in the third iteration and finally both of them accept. The negotiation terminates with success.

5   Discussion

We have developed a framework in which agents can negotiate their privacy constraints. Our framework can be used by any agent that adheres to the agent skeleton. Our particular agent makes use of ontologies and semantic rules to reason on its user's privacy constraints. Contrary to typical negotiation frameworks, only one agent proposes offers and the other agents comment on the offer by approving or giving reasons for disapproving. Agent itself collects the reasons and creates a new offer if necessary.

Particularly, in various negotiation frameworks, modeling the opponent and learning from that has been an important concept. Aydogan and Yolum have shown how agents can negotiate on service descriptions and learn each other's service preferences over interactions [2]. In this work, we have not studied learning aspects. However, the general framework is suitable for building learning agents. Particularly, since the history of reasons for rejecting a post request is being maintained, a learning algorithm can generalize over the reasons. The idea of privacy negotiation has been briefly studied in the context of e-commerce and web-based applications. Bennicke *et al.* develop an add-on to P3P to negotiate service properties of a website [3]. Their negotiation scheme is based on users' predefined demands without any semantics as we have done here. Similarly, Walker *et al.* propose a privacy negotiation protocol to increase the flexibility of P3P through privacy negotiations [13]. Their negotiation protocol is expected to terminate within a finite number of iterations and produce Pareto-optimal policies that are fair to both the client and the server. However, their understanding of privacy is limited to P3P and do not consider context-based and network-based privacy aspects.

The following three systems are important to consider regarding our purposes. Primma-Viewer [14] is a privacy-aware content sharing application running on Facebook. It is a collaborative approach where privacy policies are managed by co-owners of the shared content. Briefly, a user uploads a content and initializes a privacy policy where she defines who can access this content. She can invite others to edit the privacy policy together. Facebook gives its users the ability to report a specific post shared by others. The user can simply ask her friend to take the post down or provide more information about why she would like to report the post by selecting predefined text (e.g., "I am in this photo and I do not like it") or messaging her friend by using her own words. Face-Block [8] is a tool that converts regular pictures taken by a Google Glass into privacy-aware pictures. For this, a user specifies context-based privacy policies that include the conditions under which her face should be blurred. These policies are automatically shared with Glass users who enforce the received policies before publishing a picture. Table 4 compares our approach with these three systems, namely Primma-Viewer [14], Facebook, and FaceBlock [8] using the four criteria we defined in Sect. 1.

Table 4. Comparison of privacy criteria

Software	Automation	Fairness	Concealment	Protection
PRINEGO	✓	✓	✓	✓
Primma-Viewer	✗	✓	✗	✓
Facebook	✗	✗	✓	✗
FaceBlock	✓	✗	✗	✓

- *Automation* refers to the fact that the system actually acts on behalf of the user to align privacy constraints. Since our approach is agent-based, it is automated. The agent negotiates the users' constraints and reaches a conclusion on behalf of the user. Primma-Viewer allows users themselves to script a joint policy collaboratively; hence it is not automated. Facebook allows users to report conflicts and deals with them on individual basis. Again, the interactions are done by the users. FaceBlock, on the other hand, acts on behalf of a user to detect potential users that can take pictures and interacts accordingly. Hence, their solution is automated.
- *Fairness* refers to how satisfied users are with an agreement. For example, if a user has to remove an entire post because another user does not like it, it is not fair to the initial person. Accordingly, our approach tries to negotiate the details of the post so that only the relevant constraints are resolved. Thus, we say it is fair. In the same spirit, Primma-Viewer allows user to put together a policy so that various constraints are resolved cooperatively. In Facebook, however, the only option is to request a post to be taken off all together without worrying about what details are actually violating the user's privacy. FaceBlock is more fair than Facebook in that the content is not taken off the system but the person that has privacy concerns is blurred in the picture. However, still the user might be unhappy about other details of the picture, such as its date or context, but that cannot be specified or resolved. Hence, we consider both Facebook and FaceBlock as unfair.
- *Concealment* refers to whether the users' privacy constraints become known by other users. A user might want to say in what ways a post violates her privacy but may not want to express more. Our approach enables this by allowing users to respond to post requests with reasons. For example, a user might say that she does not want Alice to see her pictures but does not need to say why this is the case. In this respect, our approach conceals the actual

privacy concerns. Contrast this with Primma-Viewer, where the users together construct policies. In that case, all users will be aware of privacy rules that are important for each one of them; thus, the privacy constraints will not be concealed. In Facebook, everyone's privacy setting is known only by the individual and thus conceals privacy concerns from others. In FaceBlock, the privacy rules are sent explicitly to the user that is planning to put up a picture. The user evaluates the privacy rules of others to decide if the picture would bother them. This contrasts with our approach where privacy rules are private and only evaluated by the owner of the rules.

– *Protection* refers to when a system deals with privacy violations. In our work, we resolve privacy concerns before a content is posted; hence privacy is protected up front. The same holds for Primma-Viewer since the joint policy is created up front, whenever a content is posted, it will respect everyone's privacy policy. In Facebook, this is the opposite. After a content is posted, if it violates someone's privacy, that individual complains and requests it to be taken off. At this time, many people might have already seen the content. Finally, FaceBlock enforces the privacy rules before the content is put up; hence protects the privacy.

In our future work, we first plan to have adapting agents that can learn the privacy sensitivities of other agents, in terms of contexts or individuals so that the negotiations can be handled faster. For example, assume an agent never wants her protest pictures to be shown online. If her friends' agents learn about this, they can stop tagging her in such pictures. Second, we want to improve our revise algorithm to generate better offers by the use of argumentation approach so that agents can convince each other to negotiate on a content.

References

1. Andrews, L.: I Know Who You Are and I Saw What You Did: Social Networks and the Death of Privacy. Simon and Schuster, New York (2012)
2. Aydogan, R., Yolum, P.: Learning opponent's preferences for effective negotiation: an approach based on concept learning. Auton. Agents Multi-Agent Syst. **24**(1), 104–140 (2012)
3. Bennicke, M., Langendorfer, P.: Towards automatic negotiation of privacy contracts for internet services. In: The 11th IEEE International Conference on Networks, ICON2003, pp. 319–324. IEEE (2003)
4. Horridge, M., Bechhofer, S.: The OWL API: A Java API for OWL ontologies. Semant. Web **2**(1), 11–21 (2011)
5. Horrocks, I., Patel-Schneider, P.F., Boley, H., Tabet, S., Grosof, B., Dean, M., et al.: SWRL: A semantic web rule language combining OWL and RuleML. W3C Member Submission **21**, 79 (2004)
6. Jennings, N.R., Faratin, A.R.L.P., Parsons, S., Sierra, C., Wooldridge, M.: Automated negotiation: Prospects, methods and challenges. Int. J. Group Decis. Negot. **10**(2), 199–215 (2001)
7. Kafalı, O., Günay, A., Yolum, P.: Detecting and predicting privacy violations in online social networks. Distrib. Parallel Databases **32**(1), 161–190 (2014)

8. Pappachan, P., Yus, R., Das, P.K., Finin, T., Mena, E., Joshi, A.: A semantic context-aware privacy model for FaceBlock. In: Second International Workshop on Society, Privacy and the Semantic Web - Policy and Technology (PrivOn), Riva del Garda, Italy, October 2014
9. Schmidt, A., Beigl, M., Gellersen, H.-W.: There is more to context than location. Comput. Graph. **23**(6), 893–901 (1999)
10. Sirin, E., Parsia, B., Grau, B.C., Kalyanpur, A., Katz, Y.: Pellet: A practical OWL-DL reasoner. Web Semant. Sci. Serv. Agents World Wide Web **5**(2), 51–53 (2007)
11. Squicciarini, A.C., Xu, H., Zhang, X.L.: Cope: Enabling collaborative privacy management in online social networks. J. Am. Soc. Inf. Sci. Technol. **62**(3), 521–534 (2011)
12. Stewart, M.G.: How giant websites design for you (and a billion others, too). TED Talk (2014)
13. Walker, D.D., Mercer, E.G., Seamons, K.E.: Or best offer: A privacy policy negotiation protocol. In: IEEE Workshop on Policies for Distributed Systems and Networks, POLICY 2008, pp. 173–180. IEEE (2008)
14. Wishart, R., Corapi, D., Marinovic, S., Sloman, M.: Collaborative privacy policy authoring in a social networking context. In: Proceedings of the 2010 IEEE International Symposium on Policies for Distributed Systems and Networks (POLICY), pp. 1–8, Washington, DC, USA. IEEE Computer Society (2010)

Agent-Based Modeling of Resource Allocation in Software Projects Based on Personality and Skill

Mehdi Farhangian[✉], Martin Purvis, Maryam Purvis,
and Tony Bastin Roy Savarimuthu

Information Science, University of Otago, Otago, New Zealand
mehdi.farhangian@otago.ac.nz

Abstract. The success or failure of software development group work depends on the group members' personalities, as well as their skills in performing various tasks associated with the project. Moreover, in the reality, tasks have a dynamic nature and their requirements change over time. Therefore, the effect of task dynamics on the teamwork must be taken into consideration. To do so, after describing a general approach to select effective team members based on their personalities and skills, we consider as an example a comparative multi-agent simulation study contrasting two different sample strategies that managers could use to select team members: by minimizing team over-competency and by minimizing team under-competency. Based on the simulation results, we drive a set of propositions about the conditions under which there are and are not performance benefits from employing a particular strategy for task allocation. Also, we propose a simulation environment that could provide a low cost tool for managers and researchers to gain better insights about effectiveness of different task allocation strategies and employees with different attributes in dynamic environments.

Keywords: Software teams · Personality · Skill · Task allocation · Dynamic tasks · Agent-based simulation · Team management

1 Introduction

Teamwork is an essential aspect of organizational work, and there have been a number of investigations into team composition and personality [1, 2]. However, these studies have produced inconsistent results mainly because of two main constraints: firstly, they mostly consider the individual aspects of employees without fully covering group factors such as cohesion, conflict, team structure and coordination. Secondly they have not considered the dynamic nature of the task in conjunction with member personalities. In reality, various aspects of task dynamics such as changes in the task requirements or interdependency level for each task affect the team effectiveness.

Wood [3] argued changes in the complexity of tasks have an effect on the relationship between task inputs and products. Zoethout et al. [4] studied the influence on task variety on the behavior of specialists and generalists. Jiang et al. [5] examined how the change in task requirement dynamically affects individual behavior. In these

© Springer International Publishing Switzerland 2015
F. Koch et al. (Eds.): CARE-MFSC 2015, CCIS 541, pp. 130–146, 2015.
DOI: 10.1007/978-3-319-24804-2_9

studies, the relationship between managers' strategies in team formation and changes in the task requirement on the team performance is not fully covered.

Regarding these issues, in this paper, we have two main contributions. Firstly, by reviewing and applying relevant literatures, a team formation model is developed to calculate team formation performance based on personality composition and skills competency.

Secondly, we examine the relationship between the dynamic nature of tasks and managers' strategies for task allocation by using computer simulation. We model the evolution of task performance in terms of two types of parameters: task requirements and the personality distribution of employees. The simulation results can support managers' decision-making with respect to task allocation. Moreover, the effectiveness of employees with different personality in different strategy is analyzed.

The outline of the paper is organized as follows: Firstly, based on reviewing the literature, we develop a computational model to evaluate the performance of software project teams based on skill competency in conjunction with personality composition of teams. The conceptual foundations and formal considerations of task allocation mechanisms are described. To demonstrate the application of the model, simulation studies are then presented. The simulation outputs compare two task allocation models in different tasks with different level of changes in their requirements. In addition, we study the relationship between employees' personality and different task allocation strategies.

2 Team Formation Mechanism in Software Projects

In order to make rules for forming software project teams, several studies that have tried to incorporate social psychology factors for building teams [1, 6]. Among them, there is widespread recognition of the role of Myers-Briggs Type Indicator (MBTI) [7] and Belbin Team Roles (BTRs) [8] with respect to team performance.

2.1 MBTI and Belbin Team Roles

Myers [9] extended Jung psychological type [10], and it has evolved into what is now referred to as the Myers-Briggs Type Indicator (MBTI) scheme [7], which has four "dimensions" of human personality:

- Introversion vs. Extraversion (I-E) – the degree to which one faces the outer social world or keeps more to himself or herself.
- iNtuition vs. Sensing (N-S)– the degree to which one gathers information that is in concrete, objective form or is more abstract and understood according to one's inner compass.
- Thinking vs. Feeling (T-F) – the degree to which one makes decisions based on logic and demonstrable rationality or is more empathic and attempts to see things from given social perspectives.
- Perceiving vs. Judging (P-J) – the degree to which one wants to come to quick, categorical decisions or is more inclined to withhold judgment for the time being.

Belbin [8] introduced a theory about the roles of individuals in teams. In each team, every member has a role that might affect the performance of the team. In an early publication, eight team roles were identified: Chairman, Shaper, Plant, Monitor-Evaluator, Company Worker, Resource Investigator, Team Worker, and Completer-Finisher [4]. Later he added a ninth role, Specialist and renamed the Chairman to Coordinator and the Company Worker to Implementer [11]. Other researchers then raised the possibility that the relationship could be found between the MBTI. These roles are explained in Table 1 [8].

Table 1. Belbin Roles

Team Role	Contribution	Allowable weakness
Plant	Creative	Ignores incidentals
Resource investigator	Outgoing, enthusiastic	Over-optimistic
Coordinator	Mature, confidant	Can be seen as manipulative
Shaper	Challenging, dynamic	Prone to provocation
Monitor evaluator	Sober, strategic	Lacks drive to inspire others
Team worker	Cooperative	Indecisive in crunch situations
Implementer	Practical	Somewhat inflexible
Completer	Painstaking	Inclined to worry unduly
Specialist	Single-minded	Contributes only on a narrow front

Personality profiles and Belbin Team Roles (BTRs) suggest that personality and role tendencies are not independent [12]. Stevens and Henry [13] tried to map these two instruments [14], Stevens [13] noticed that there is a different distribution of both BTRs and MBTI and from this distribution the personality related to the team roles could be determined, and Schoenhoff [15] continued this work by using a larger sample.

Myers also introduced a theory, namely MTR-i [16], which incorporates the idea of team roles, and he claimed people with different personalities are likely to have specifically correlated roles in a team. Table 2 compares the results of different studies (where X means no relationship between personality and Belbin role is found). The rightmost column of Table 2 indicates the degree of commonality among the other four studies. We designate the agreement points for that rightmost column if, for a given Belbin role, at least two of the studies agree on an MBTI personality dimension for that role. Also, these agreement points seem to be in relative accord with the Keirsey study of temperaments [17].

3 Performance Calculation Model

In this paper, we formulate a performance computation mechanism for software development projects by taking into consideration employees' personalities and skills. The motivation for the computational model is based on the previous findings and from both MBTI and BTR studies.

Table 2. Studies about the relationship of personality and BTRs

Belbin roles [8]	Henley report [18]	Stevens report [13]	Schoenhoff report [15]	MTR-I [16]	Agreement points
Coordinator	EXXX	XSXX	ENFP	ESFP/ESTP	EXFP
Shaper	EXXX	EXXX	XSTJ	ESFP/ESTP	ESTX
Plant	IXTX	XNTP	INTJ	INTJ/INFJ	INTJ
Monitor evaluator	IXTX	XXXX	ISXJ	ISTJ/ISFJ	ISTJ
Implementer	XXXX	XSXJ	ISXJ	XXXX	XSXJ
Resource investigator	EXXX	EXXP	ENFJ	ENTP/ENFP	ENFP
Team worker	EXXX	XXXX	ISTJ	ESFJ/ENFJ	ESXJ
Completer	IXXX	XSXJ	ISTJ	XXXX	ISXJ
Specialist	XXXX	XXXX	XXXX	ISTP/INTP	XXXX

Belbin suggests two main factors for forming a team: dyadic relationships of team members and competency of team members in the tasks [8]. In this connection, we describe a formal model that represents the assignment of people to the software projects and which reflects the literature about team formation. Managers calculate the performance of each team composition and select the best one for their task. The general formula for calculation of team performance is expressed as follows.

$$Performance = \text{Personality\_Composition} * Competency \tag{1}$$

$$\begin{aligned} Personality\_composition = (&c1 * Matching\_Personality + c2 * Matching\_roles + \\ & c3 * Creativity\_capability + c4 * Urgency\_capability + c5 * Sociality\_capability + \\ & c6 * Complexity\_capability + c7 * \text{Belbin}\_Creativity\_capability + c8 * \\ & \text{Belbin}\_Urgency\_capability + c9 * \text{Belbin}\_Sociality\_capability + c10 * \\ & \text{Belbin}\_Complexity\_capability) \end{aligned} \tag{2}$$

To express this more compactly, we can write this as

$$\begin{aligned} Performance = (&c_1 * Pm + c_2 * Rm + c_3 * Cr + c_4 * Um + c_5 * So + c_6 * Co + c_7 * Bcr + \\ & c_8 * Bum + c_9 * Bso + c_{10} * Bco) * c_{11} * C \end{aligned} \tag{3}$$

The various parameters, such as *Matching_personality, (Pm), Matching_roles (Mr), ..., C (Competency)* are explained and formulated in the next sections. These variables are numerical values that can be uniformly taken to be measured along some scale, such 0 to 1 and each one explained in the following sections. The identifiers $c1, ..., c11$ are coefficients that can be adjusted for fitting empirical measurements. In this formulation for team performance, we have considered the factors that were most prevalent from our literature survey. Further variables of our model are described as follows:

m: the number of skills required for tasks
n: the number of employees for each team
R_k: the skills requirement vector for task k. Thus $R_k = [R_{k1}, R_{k2} ..., R_{km}]$
im: an index identifier indicating the most important skill
$R_k[im]$: the skill requirement of the most important skill for the task k
S_i: the skills vector of employee i. $S_i = [S_{i1}, S_{i2}, ..., S_{im}]$

These parameters are based on our literature survey, and we provide further descriptions of these factors in the following. We describe skill competency and personality composition that are mentioned in formula 1 as follows:

3.1   Skill Competency of Team Members (C)

An important factor is the competency or skills of the team. We calculate the competency for each skill by dividing the skill of an employee by the skill requirements for the task. The overall team competency is the sum of all the team members' competencies for each skill.

In practice, managers have various preferences for task allocation. The standard approach is to find the minimal difference between the skills of employees and the task demands, and it is used in different ways in the literature for personnel selection [25]. However, existing methods have not considered a positive and negative gap values in connection with the differences. In our model, we propose a similarity measure such that a positive gap value is considered as over-competency and a negative value is considered as under-competency. These two methods are presented as two different task allocation strategies. For each strategy, the manager will calculate a utility skill competency of team and choose teams with the highest value.

3.1.1   Minimizing Under-Competency

In this method, the main purpose of the manager is minimizing under-competency in assigning the task to the employees. They try to choose the best combinations of employees who have the least under competency in their skill. So they calculate the utility of teams based on the following formula. Where C_{il} represents the competency of employees in the skill in this mechanism, R_l represents the skill requirement of task l, and S_{il} represents the skill of employee i in task l.

$$C_{il} = 1 - \max(0, (R_l - S_{il})/R_l) \tag{4}$$

3.1.2   Minimizing Over-Competency

In this method, the main purpose of the manager is minimizing over-competency in assigning the task to the employees. They try to choose the best combinations of employees who have the least over competency in their skill. So they calculate the utility of teams based on the following formula. C_{il} represents the competency of employees in skill in this mechanism.

$$C_{il} = \begin{cases} 1 - \frac{(S_{il} - R_l)}{R_l} & if \quad S_{il} - R_l \geq 0 \\ 1 - \frac{(R_l - S_{il})}{R_l} & if \quad S_{il} - R_j < 0 \end{cases} \tag{5}$$

3.2 Personality Composition

The first ten factors in Formula (3) are related to the personalities of team members. We measure the goodness of team composition by factors such as matching their Belbin's roles, matching their MBTI Personality, team creativity, the MBTI capability of team to dealing with task requirements such as creativity, urgency, sociality, and task complexity, and the Belbin capability of the team to deal with task requirements such as creativity, urgency, sociality and task complexity. Each factor is described as follows:

Matching_roles (Rm): Matching roles represents the degree to which Belbin roles are suitably matched. All the people have a primary natural team role that affects their behavior with each other. The interactive relationships of team members influence the team environment and performance. For example, if someone is aggressive towards someone, the recipient may respond by being diplomatic or by having a significant clash with the aggressor. Belbin's study shows this interpersonal relationship and what kind of people have likely conflict with each other and what kind of people tend to work well with each other. In Table 3, we summarize these interpersonal relationships from Belbin's work [8].

Table 3. Belbin's roles

Role	Suitable peer	Unsuitable peer
Shaper	Resources investigators	Plant
Specialist	Implementers, team workers	Plant
Monitor evaluator	Coordinators, implementers	Completers, other monitor evaluators
Completer	Implementers	Resource investigators
Implementer	Coordinators, monitor evaluators, resource investigators, completers and specialists	Other implementers and plants
Resource investigator	Implementers and team workers	Completers and specialists
Coordinator	Implementers and team workers	Shapers
Team worker	Other team workers and plants	Shapers

On the basis of these relationships, we formulate the index *Rm* as an indication of relationship compatibility:

$$Rm_b = \frac{(Ps_b - Pu_b)}{\max[Ps_b, Pu_b]} \tag{6}$$

Where Rm_b is the degree of matching of peers' roles in team b, Ps_b is the number of suitable roles in the team, and Pu_b is the number of unsuitable roles in the team.

Matching_index (Pm): Matching-index (Pm) represents the degree to which personalities, as measured by MBTI type, are matched. We base this on studies about the effect of personality composition of a team. As with Belbin's roles, some personalities do not get along well with each other, so it can be important to configure team personalities appropriately. We have surveyed the literature concerning personality composition of teams, and Table 4 shows the relationship conflicts across MBTI personality types. These assumptions are based on [19–23].

Table 4. Relationships of MBTI personality dimensions

	T	F			J	P			E	I			S	N
T	0	+		J	+	-		E	-	0		S	0	+
F	+	0		P	-	+		I	0	0		N	+	0

Note that in the table, '+' means that there is a positive effect, '−'means there is a negative effect, and '0' means that there is no effect.

It has been found, for example that two extraverted people working together can be problematic because they can be dominant and assertive towards each other. Additionally, it has been found that Sensing and iNtution types can be useful to each other, as well as Feeling and Thinking. People who differ across the Judging and Perceiving dimension tend to frustrate each other, but people at the same end of the Judging or Perceiving scales have similar interests and can understand and predict each other's behavior.

For each of the four MBTI personality dimensions, we established a scale between 0 and 100 and assigned values for each employee.

- Introverted/Extraverted: (range 0–50 → *Introverted*; 50–100 → *Extraverted*).
- Intuitive/Sensing: (range 0–50 → *Intuitive; 50–100 → Sensor*),
- Thinking/Feeling: (range 0–50 → *Feeler; 50–100 → Thinker*),
- Perceiving/Judging: (range 0–50 → *Perceiver*; 50–100 → *Judgers*).

Using these parameters, we construct the final score for matching personality as:

$$EE_{ij} = \begin{cases} -\frac{(Extraverted_i + Extraverted_j)}{200} & \text{if } Extraverted_i > 50 \text{ and } Extraverted_j > 50 \\ 0 & \text{otherwise} \end{cases} \tag{7}$$

$$SN_{ij} = \begin{cases} \frac{\left[\left(Sensing_i - sensing_j\right)\right]}{100} & \text{if } sensing_i > 50 \text{ and } sensing_j \leq 50 \\ & \text{or } sensing_i \leq 50 \text{ and } sensing_j > 50 \\ 0 & \text{otherwise} \end{cases} \quad (8)$$

$$TF_{ij} = \begin{cases} \frac{\left[\left(Feeling_i - Feeling_j\right)\right]}{100} & \text{if } Feeling_i > 50 \text{ and } Feeling_j \leq 50 \\ & \text{or } Feeling_i \leq 50 \text{ and } Feeling_j > 50 \\ 0 & \text{otherwise} \end{cases} \quad (9)$$

$$JP_{ij} = \begin{cases} \frac{\left(Judging_i + Judging_j\right)}{200} & \text{if } Judging_i > 50 \text{ and } Judging_j > 50 \\ \frac{\left(100 - Judging_i\right) + \left(100 - Judging_j\right)}{200} & \text{if } Judging_i \leq 50 \text{ and } Judging_j \leq 50 \\ -\frac{\left[\left(Judging_i - Judging_j\right)\right]}{100} & \text{if } Judging_i > 50 \text{ and } Hudging_j \leq 50 \text{ or} \\ & Judging_i \leq 50 \text{ and } Judging_j > 50 \end{cases} \quad (10)$$

Using these parameters, we construct the final score for matching personality between employee i and employee j.

$$Rp_{ij} = \frac{\left(EE_{ij} + SN_{ij} + TF_{ij} + JP_{ij}\right)}{4} \quad (11)$$

The matching personality of a team expressed as follows:

$$Pm_b = \frac{\left(\sum_{i=1}^{n} \sum_{j=1}^{n} Rp_{ij}\right)}{n} \quad (12)$$

In the above, EE_{ij} represents the dyadic effect of the Extraverted-Introverted dimension (in this case introversion has no effect), SN_{ij} represent the dyadic effect of the Sensing-Intuition dimension, TF_{ij} represents the dyadic effect of the Thinking-Feeling dimension, and JP_{ij} represents the dyadic effect of the Judging-Perceiving dimension. Pm_b indicates the matching personality of team b.

So far, we have just considered how personalities and roles match with each other, but we must also take into consideration how they match up with the task types. To operationalize this, we consider various tasks to have different levels with respect to (a) required creativity, (b) urgency, (c) required social interaction, and (d) complexity. Each of these categories is discussed further below. In this connection, we use two additional indicators that are useful for these considerations [24]:

- Team Personality Elevation (TPE): a team's mean level for given personality trait.
- Team Personality Diversity (TPD): the variance with respect to a personality trait

Creativity (Cr): For tasks requiring a high level of creativity, teams composed of differing attitude tendencies are believed to perform better than teams of like-minded people [24]. So, here we assume high heterogeneity (high TPD) in the four personality dimensions will lead to creativity. Moreover, the creativity of individuals is related to their Intuition level [21]. So, in addition to a high TPD in all four personality dimensions, we also assume that high TPE in Intuition has positive effects on creativity. In the following expressions, Cri_b is the combined team index for creativity, and Crr_k is the required creativity for the task.

$$Cr_b = (TPE \text{ of } Intuition + mean \text{ of } TPDs)/n * 100) \tag{13}$$

$$Cri_b = \begin{cases} Cr_b/Crr_k & \text{if } Crr_k - Cr_b \geq 0 \\ 1 & \text{if } Cr_b - Crr_k > 0 \end{cases} \tag{14}$$

Urgency (Um): When time is important, Perceiver types, who need freedom for their actions, are less likely to be successful. In contrast, Judgers relish getting in on the closure of a task, and so they can have a positive effect on tasks with time pressure. As a result, we believe that a high TPE in Judging has a positive effect in performing urgent tasks.

$$Um_b = TPE \text{ in } Judgers/n * 100 \tag{15}$$

$$Umi_b = \begin{cases} Um_b/Umr_k & \text{if } Umr_k - Um_b \geq 0 \\ 1 & \text{if } Um_b - Umr_k > 0 \end{cases} \tag{16}$$

Umi_b is the combined team score (index) for Urgency, and Umr_k is the required Urgency for the task.

Sociality (So): For tasks involving many social interactions, extraverted individuals can help the team. Therefore, we assume a high TPE in Extraversion has a positive effect in performing these tasks.

$$So_b = TPE \text{ in } extraverted/n * 100 \tag{17}$$

$$Soi = \begin{cases} So_b/Sor_k & \text{if } Sor_k - So_b \geq 0 \\ 1 & \text{if } So_b - Sor_k > 0 \end{cases} \tag{18}$$

Soi_b is the combined team is score for Sociality, and Sor_k is the required sociality for the task.

Complexity (Co): When the complexity of a task is high, a rational and scientific mind that is characteristic of thinking types can be useful. As a result, we expect a high TPE in Thinking will have a positive effect in performing these tasks.

$$Co_b = TPE \text{ in } Thinking/n * 100 \tag{19}$$

$$Coi_b = \begin{cases} Co_b/Cor_k & \text{if } Cor_k - Co_b \geq 0 \\ 1 & Co_b - Cor_k > 0 \end{cases} \tag{20}$$

Coi_b is the combined team score for complexity, and Cor_k is complexity of the task.

In addition to the above eight indicators, we assume that some roles are crucial for some tasks, so we have introduced the following constraints based on Belbin's findings [11]. Having

- at least one Plant is essential in teams with a high creativity requirement.
- at least one Completer is essential in teams with a high urgency requirement.
- at least one Evaluator is essential in teams with a high complexity requirement.
- at least one Resource Investigator is essential in teams with a high complexity requirement.

Regarding the above mentioned rules and constraints which are extracted from various from literatures on team performance and personality, we develop and agent-based model for task allocation. We have used optimization and filtering algorithms to compute the utility of all the combinations. The system searches for all the possible combinations of a team and calculates the highest valued coalition. The system then assigns tasks to the employees to maximize the utility of the system. The following algorithm is used for the highest valued coalition U^*.

1.   *V ← set all the possible task order*
2.   **For each** *order*
3.   *C ← set all the combinations of employees*
4. U_j *← Set the utility of each employee combination with task j*
             based to mangeres' task allocation strategy
5. U_j^* *← set the maximum of U_j*
6. U_{max} *← argmax U_i^**
7.   *if employee$_i$ ∈ U_{max} and*
           it doesn't violate any constraints about task j
8.   **then** *compute the effeciency of U_{max} with formula* (3)
9.   *Delete all the combinations in C that contain employee i*
10.   *if C is not empty*
11.   **then** *goto 2*
12.   *Calculate: Z ← $\sum U_{max}$*
13. *Choose Max Z*
14. **Return**

4   Simulation and Results Analysis

In order to explore the effect of task dynamics of our model on the proposed task allocation mechanism, we conducted some simulation experiments on the NetLogo platform [26].

In this model that is depicted in Fig. 1, the dynamic tasks are characterized by changing the requirements of tasks. In the reality, managers have to reschedule their projects because of new requirements for tasks. Rescheduling has some cost since it takes time for new member to be familiar with the new tasks, and it causes some dissatisfaction for those who leave the task. In each time step, with a certain probability,

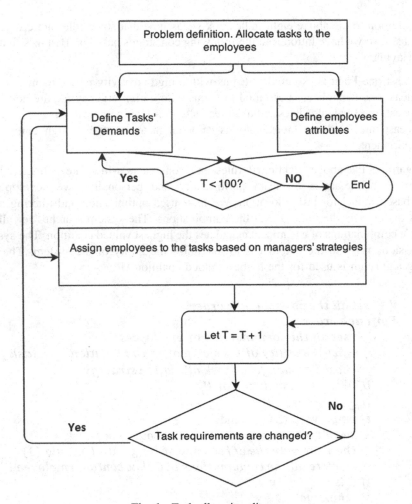

Fig. 1. Task allocation diagram

the requirements of one skill increase and managers select the best team for this task. So, in each time step managers calculate the payoff of changing teams, and if this payoff is positive, they change the team. This payoff is calculated by the following formula:

$$Payoff = (\sum_{i=1}^{n} C_{inew} - C_{e}t) - \sum_{i=1}^{n} C_{icurrent} \tag{25}$$

Where C_{inew} and $C_{icurrent}$ represents the competency of new and current team members respectively.

The cost of changing a team is a constant number and is indicated by C_e. The cost, of changing the current team, is formulated by $C_e t$. This cost is related to the time that has elapsed from the starting point of the project. As a result, the skill competency of team is calculated according to the following formula.

$$Ds_{bk} = \begin{cases} \sum_{i=1}^{n} C_{inew} - C_e t & if \ (\sum_{i=1}^{n} C_{inew} - C_e t) > \sum_{i=1}^{n} C_{icurrent} \\ \sum_{i=1}^{n} C_{icurrent} & if \ (\sum_{i=1}^{n} C_{inew} - C_e t) \leq \sum_{i=1}^{n} C_{icurrent} \end{cases} \qquad (26)$$

$$Performance(t) = Ds_{bk} * Personality\_Composition_b * t \qquad (27)$$

Where $Performance(t)$ indicates the performance of team in time t, $Composition_b$ indicates the personality composition of team b and calculated as presented in the formula 2. C_{il} indicates the competency of agent i in discipline l and C_b presents the competency of members in task b.

The experiments, we compare the performances of two managers who assign the employees to the tasks. In order to calculate the competency C_k, the manager with "Minimizing Under-competency" strategy uses Formula (4) and the manager with "Minimizing Over-competency" strategy uses Formula (5).

The simulation environment could provide a low cost simulation tool for the managers and researchers to investigate the impact of employees' and tasks' attributes and dynamism of environment and also their task allocation strategies on team performance. A schematic representation of the proposed tool is presented in Fig. 2.

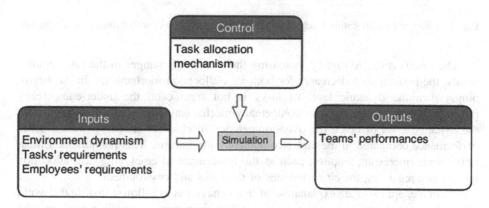

Fig. 2. Input-output and control parameter of proposed simulation tool

In the initial settings, the environment had 12 employees and four tasks. Both tasks and employees have some initial properties. In this connection, a task role is assigned to each person, and the choice for this role is guided by the personality information from Agreement Points (right-hand-most) column of Table 1. Values between 0 and 10 are assigned to the employees (these skills levels are assigned according to a normal distribution with a standard deviation of 3). In addition, specific task attributes are assigned to the task, such as the required level of creativity, social interactions, complexity, and urgency. A number between 0 and 100 is assigned to each such task attribute. Three skills are allocated to the task representing the skills that are required, and a number between 0 and 10 represents the required skill level. For the sake of simplicity, we assume that all teams comprise a small number (three) of employees. Also in the simulation settings, number 1 is assigned to $C1, \ldots, C11$.

The results of simulation experiments are summarized in Fig. 3. It compares the simulation results of the two task allocation methods with different probabilities of increasing the task requirements in each time step. The results are averaged over 100 runs of the model.

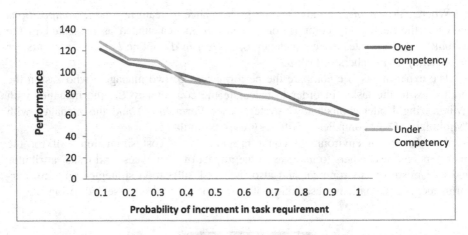

Fig. 3. Effect of task allocation mechanism on performance of tasks with dynamic requirements.

The results revealed that by increasing the chance of changes in the task requirements, the performance decreases for both task allocation mechanisms. In the beginning, when the dynamic level of tasks is not significant, the under-competency mechanism outperforms the over-competency mechanism. However, after increases in the dynamic level of tasks, the over-competency mechanism ended up with a better performance compared to the under-competency mechanism. This phenomenon illustrates some interesting features, such as the importance of employing task allocation mechanism regarding the characteristics of the tasks and environment.

A simple, approximate explanation of this behavior is as follows: first, in the world when the probability of changes in the task requirement is small, managers who minimize over competency are more likely to make mistakes. For instance, among two employees that one is overqualified, and another one is underqualified over-competency managers might choose the underqualified one that will result in the poor performance. When this probability increases, the managers who minimize under competency make more mistakes. It occurs since the employee selection among some overqualified employees is a random process for these managers. For performing the next projects, they might want to assign these overqualified workers to tasks that are really required. This phenomenon occurs more in a dynamic environment and results in some costs for the under-competency managers.

4.1 Relationship of Personality and Performance

In order to understand the relationship between personality and dynamic tasks, we conducted further simulation experiments. In the previous experiments, we assigned random personalities to the employees. In contrast, in these experiments, some scenarios are evaluated with respect to various personality configurations. We examined the performances of members with different distributions of personality when the probability of changing the requirements of the task in each time-step is 0.3. In other words, we are interested in examining whether a task allocation mechanism has any advantages over another one for a particular personality distribution. In order to assess the robustness of each personality distribution and qualify the certainty of predictions arising from experiments, we used a *one-at-a-time* uncertainty analysis technique, the Vargha-Delaney A-test [27], which is a non-parametric effect magnitude test, to determine when a parameter adjustment has resulted in a significant change in simulation behaviour from the baseline. The test compares two population distributions and returns a value in the range [0.0, 1.0] that represents the probability that a randomly chosen sample taken from the population A is larger than a randomly chosen sample from population B. Table 5 show how the A-test scores relate to various magnitudes of differences between two populations. For this simulation test baseline behaviour is required, and we used here the personality distribution when personalities are assigned randomly.

Table 5. The magnitude effect by A-test score

Differences	Large	Medium	Small	None	Small	Medium	Large
A score	0.29	0.36	0.44	0.50	0.56	0.64	0.71

In our experiments, we have 20 scenarios; each scenario represented a different personality distribution, and the results are summarized in Table 6. In each scenario, we measure the probability that the under-competency mechanism performs better than the over-competency mechanism.

Table 6. The effects of different personality distributions in the comparison of the over-competency strategy with under competency strategy.

	I-E	N-S	T-F	P-J
0 %–100 %	0.391	0.53	0.578	0.312
25 %–75 %	0.432	0.522	0.504	0.366
50 %–50 %	0.476	0.513	0.451	0.397
75 %–25 %	0.493	0.43	0.424	0.492
100 %–0 %	0.545	0.37	0.405	0.581

For instance, the first number in the left-top of the Table 6 is 0.391. This number means in the case that 0 % of employees are introverted, and 100 % are extraverted the probability that under the competency mechanism performs better than over competency is 0.391. We found that the magnitude of the performance advantages depends not only on the personality distribution, but also on task allocation strategy. In most of the cases (different distribution of personality), there were none or only a small magnitude effect measured by the A-Test score between task allocation mechanisms. In most of the scenarios, the probability, of having a better performance with under-competency mechanism is slightly better than the other task allocation mechanism. However, we observed in some scenarios the over-competency mechanism outperformed the under-competency mechanism with a medium magnitude effect. For example, when 100 % of the employees have Judging type, the A-score is 0.581, which means the probability that the over-competency performs better than under-competency is 0.581. In general, the over-competency mechanism had slightly better performances in cases when the majority of employees were Feeling or Perceiving or Sensing or Extraverted.

These observations are interesting and can be explained approximately. For instance, when the majority of employees are Extraverted, minimizing over competency more likely save some of the capability of the organization for the next projects with a high sociality requirement.

5   Conclusion

In this paper, we have proposed a computational model, parameterized on the basis of reports in the academic literature, for measuring the performance of software teams considering their personality composition and skill competency. Based on this concept, we examine the effect of managers' strategies for task allocation on team performance when they are dealing with dynamic tasks. We ran agent-based simulations and designed various scenarios with different degrees of dynamic level. We studied whether a resource allocation strategy leads to performance advantages with respect to dynamic tasks. We also examined whether different personality distributions have an effect on two different task allocation methods. The effects of the personality distribution and the magnitudes of the impact of each personality were measured.

Based on these experiments, we drive a set of propositions about the conditions under which there are and are not performance benefits from employing a particular strategy for task allocation. Increasing the degree of changing requirements had a more adverse effect when the strategy of managers is minimizing under-competency compared to when the strategy of managers is minimizing over-competency. In addition, in most cases of the personality distribution, two strategies did not have significant differences; however, for a few scenarios some exceptions were observed.

We propose a multi-agent tool that can be used for researchers and managers to investigate the effect of their employees and task allocation strategies in a real-world environment. Our goal is to provide a comprehensive model for managers to investigate the impact and effectiveness of (1) different task allocation strategies in different dynamic environment (2) employees with different attributes in terms of personality and skill.

We wish to note here that what we are presenting here as a contribution is not so much the specific simulation results, but a modelling and simulation approach that can demonstrate interesting emergent effects based on combinations of personality and skill configuration parameterizations. This parameterization can be set for the specific contextual circumstances to examine sensitivities in this area.

We believe there is no universally successful personality configuration, and the situational forces such as organizational and cultural forces and task structures must be taken into consideration before generalizing the proposed rules between personality and performance. Regarding that, in this paper we mainly emphasize on the proposed framework for building an agent-based model and the results can vary in different domains. Researchers and managers might change rules, formulas and the constraints in the team formation mechanism section based on situational forces.

Our work would be enhanced by the availability of real data that could be used to validate the assumptions and the results. In the future, we will be gathering data concerning these tasks allocation mechanisms from groups of software engineering students undertaking group projects. Also, this system could be used to assist real managers to keep track of their task allocation activities. We intend to provide a decision-support system tool that employs our modeling approach to support managers' activities in dealing with dynamic tasks.

References

1. André, M., Baldoquín, M.G., Acuña, S.T.: Formal model for assigning human resources to teams in software projects. Inf. Softw. Technol. **53**(3), 259–275 (2011)
2. LePine, J.A., Buckman, B.R., Crawford, E.R., Methot, J.R.: A review of research on personality in teams: accounting for pathways spanning levels of theory and analysis. Hum. Resour. Manage. Rev. **21**(4), 311–330 (2011)
3. Wood, R.E.: Task complexity: definition of the construct. Organ. Behav. Hum. Decis. Process. **37**(1), 60–82 (1986)
4. Zoethout, K., Jager, W., Molleman, E.: Task dynamics in self-organising task groups: expertise, motivational, and performance differences of specialists and generalists. Auton. Agent. Multi. Agent. Syst. **16**(1), 75–94 (2007)
5. Jiang, G., Hu, B., Wang, Y.: Agent-based simulation approach to understanding the interaction between employee behavior and dynamic tasks. Simulation **87**(5), 407–422 (2010)
6. Acuña, S.T., Gómez, M., Juristo, N.: How do personality, team processes and task characteristics relate to job satisfaction and software quality? Inf. Softw. Technol. **51**(3), 627–639 (2009)
7. Myers, I.B., McCaulley, M.H.: Manual: a Guide to the Development and Use of the Myers-Briggs Type Indicator. Consulting Psychologists Press (1985)
8. Belbin, R.M.: Team Roles at Work. Routledge, London (2012)
9. Myers, I.: The Myers-Briggs Type Indicator. Consulting Psychologists Press (1962)
10. Jung, C.G.: Psychological Types: or the Psychology of Individuation. Harcourt, Brace (1921)
11. Belbin, R.M.: Management Teams: Why they Succeed or Fail. Bulletin of the British Psychological Society (1981)
12. Stevens, K.T.: The effects of roles and personality characteristics on software development team effectiveness. Dissertation, Virginia Polytechnic Institute and State University (1998)

13. Henry, S.M., Todd Stevens, K.: Using Belbin's leadership role to improve team effectiveness: an empirical investigation. J. Syst. Softw. **44**(3), 241–250 (1999)
14. Stevens, K., Henry, S.: Analysing software teams using Belbin's innovative plant role. Department of Computer and Information Science, University of Mississippi and Department of Computer Science, Virginia Tech (2002)
15. Schoenhoff, P.K.: Belbin's Company Worker, The Self-Perception Inventory, and Their Application to Software Engineering Teams. Dissertation, Virginia Polytechnic Institute and State University (2001)
16. Myers, S.: MTR-i: a new arena for team roles. Training JOURNAL-ELY- 24–29 (2002)
17. Keirsey, D.: Please understand me II: temperament, character, intelligence. Prometheus Nemesis (Del Mar, CA), p. 350 (1998)
18. Higgs, M.J.: A comparison of Myers Briggs type indicator profiles and Belbin team roles. Henley Business School, University of Reading, 21 Aug 1996
19. Chen, S.-J.G.: An integrated methodological framework for project task coordination and team organization in concurrent engineering. Concurr. Eng. 185–197 (2005)
20. Bayne, R.: The Myers-Briggs Type Indicator: a Critical Review and Practical Guide (1995)
21. Bradley, J.H., Hebert, F.J.: The effect of personality type on team performance. J. Manage. Dev. **16**(5), 337–353 (1997)
22. Culp, G., Smith, A.: Understanding psychological type to improve project team performance. J. Manage. Eng. **17**(1), 24–33 (2001)
23. Varvel, T., Adams, S.G., Pridie, S.J., Ruiz Ulloa, B.C.: Team effectiveness and individual Myers-Briggs personality dimensions. J. Manage. Eng. **20**(4), 141–146 (2004)
24. Bowers, C.A., Pharmer, J.A., Salas, E.: When member homogeneity is needed in work teams: a meta-analysis. Small Group Res. **31**(3), 305–327 (2000)
25. Canos, L., Liern, V.: Some fuzzy models for human resource management. Int. J. Technol. Policy Manage. 291–308 (2004)
26. Tisue, S., Wilensky, U.: NetLogo : design and implementation of a multi-agent modeling environment. In: Proceedings of Agent (2004)
27. Vargha, A., Delaney, H.D.: A critique and improvement of the cl common language effect size statistics of McGraw and Wong. J. Educ. Behav. Stat. **25**(2), 101–132 (2000)

On Formalizing Opportunism Based on Situation Calculus

Jieting Luo(✉), John-Jules Meyer, and Frank Dignum

Utrecht University, PO Box 80089, 3508 TB Utrecht, The Netherlands
{J.Luo,J.J.C.Meyer,F.P.M.Dignum}@uu.nl

Abstract. In social interactions, it is common for one party to possess more or better knowledge about a specific transaction than others. In this situation, parties who are more knowledgeable might perform opportunistic behavior to others, which is against others' interest thus leading to relationship deterioration. In this study, we propose formal models of opportunism, which consist of the properties knowledge asymmetry, value opposition and intention, based on situation calculus in different context settings. We illustrate our formalization through a simple example. Further study on its emergence and constraint mechanism can be carried out based on the formal models.

Keywords: Opportunism · Value · Situation calculus · Formalization

1 Introduction

Consider a common social interaction. A seller is trying to sell a cup to a buyer and it is known by the seller beforehand that the cup is actually broken (e.g. there is a crack at the bottom of the cup). The buyer buys the cup for its good appearance, but of course gets disappointed when he fills it with water. In this example, the seller earns money from the buyer by exploiting the opportunity of knowledge asymmetry about the cup, while the buyer just focuses on the appearance of the cup rather than being leaky or not. Such a social behavior intentionally performed by the seller is first proposed by economist Williamson as opportunism [1]. Opportunistic behavior commonly exists in business transactions and other types of social interactions in various forms such as deceit, lying and betraying.

Viewing individuals as agents, we may have similar problems in multi-agent system research. Interacting agents were modeled to behave in a human-like way with characteristics of autonomy, local views and decentralization [2]. When such agents possess different quantity or quality of relevant information and try to maximize their benefits, they may probably perform opportunistic behavior to others, which is against others' benefits or the norms of the system. For example, in a system with the norm of equity, an agent may hide important information to his or her peers for increasing his own payoff. The agent's behavior has negative results for other agents involved in the relationship and strongly affects the cooperative relationship once it is unveiled.

© Springer International Publishing Switzerland 2015
F. Koch et al. (Eds.): CARE-MFSC 2015, CCIS 541, pp. 147–166, 2015.
DOI: 10.1007/978-3-319-24804-2_10

In order to explore this problem, we need to have a formal model of opportunism which can be applied in any context and serve as a basic framework for future research. Logic-based formalisms are one of the alternatives for its capacity of describing and reasoning. Through the specification by logic, we can understand more clearly the elements in the definition and how they constitute this social behavior. More importantly, we can derive interesting properties that are useful for our future research. Thus, we are motivated to propose a formal specification of opportunism by mathematical logic based on our definition. We believe that such a research perspective can ease the debates about opportunism in social science. Moreover, future work on its emergence and constraint mechanism can be conducted based on our formal model, rendering our study relevant for MAS research.

In this paper, we first have a clear definition of opportunism extended from Williamson's, highlighting the key elements we need to model. Aiming at the investigation about the different judgement on opportunistic behavior, we integrate the notion of value to represent agents' preference on situations. We then formalize opportunism using the situation calculus [6,7] as our technical framework based on our extended definition. We first consider opportunistic behavior as a single action between two agents, and then extend it to multiple actions and incorporate social context in the model. Our formal models of opportunism consist of the property knowledge asymmetry, value opposition and intention, and represent how they relate to each other. We illustrate how to use our formal models through a simple example selling a broken cup.

2   Defining Opportunism with Value

In this section, we extend Williamson's definition of opportunism and suggest a more explicit one as a prelude and basis to proposing a formal model in the next section.

2.1   Definition of Opportunism

The classical definition of opportunism is offered by Williamson [1] as "self-interest seeking with guile". While this definition has been used in a large amount of research, it only mentions two attributes, self-interest and guile, explicitly, leaving other attributes for researchers to interpret from different perspectives. For example, Das defined partner opportunism as "behavior by a partner firm that is motivated to pursue its self-interest with deceit to achieve gains at the expense of the other alliance members" [3]. Even though it is elaborated enough, it has the suggestion that opportunistic individuals are meant to harm others, which cannot be derived from Williamson's definition. In this study, based on the definition of Williamson, we compare opportunistic scenarios with non-opportunistic ones, and redefine this social behavior in a more explicit way.

Opportunism is a behavior that is motivated by self-interest and takes advantage of relevant knowledge asymmetry[1] to achieve gains, regardless of the principles.

First of all, there has been reached consensus that opportunistic behavior is performed with self-interest motivation [3]. We admit that self-interested pursuit is the natural property of human beings, but opportunism is more than that: individuals with opportunistic behavior do not care about the negative effects on others.

Secondly, relevant knowledge asymmetry provides the chance to individuals to be opportunistic. Opportunistic individuals may break the contracts or the relational norms using the relevant knowledge that others do not have. It is important for opportunistic individuals to use cheating, deceit or infidelity for hiding their self-interest motive. Therefore, individuals with more relevant knowledge will have more potential for opportunistic behavior.

Thirdly, principles are ignored by opportunistic individuals. The reason to use "ignore" here is to distinguish opportunism from accidentally bringing harm to others. Opportunistic behavior is performed on purpose without any compensation to the victims. Principles can be the value of others, or the contract rules or the relational norms that are used for balancing various interests and already agreed to by a majority of the individuals.

Fourthly, even though we did not explicitly declare the result of performing opportunistic behavior in our extended definition, such a social behavior must result in gains at the expense of others. Any self-interested behavior that does not affect other individuals should not be considered as opportunism.

From the above elaboration, we have something important to keep in mind: it is not the intention of opportunistic individuals to harm others even though opportunism is deliberate with self-interest motives. The ignored principles are a specific kind of knowledge about the interest of others that cannot be considered as an intention to be opportunistic. This is one of the properties that we are going to show through our formal model of opportunism.

2.2 Integrating with Value

Based on the informal definition of opportunism, the example about hiding information that we encountered in the introduction is opportunistic behavior, since it is against others' benefits or the norms of the system. However, if hiding is not forbidden by the norm, the agent could not be said to have done anything wrong. Or if other agents agree with that agents having more important information deserve more payoff, it may not be against other agents' interest. We can see that both the system's norms and the agents' perspectives can influence the judgement of opportunism, and they are the representation of value systems

---

[1] Even though many papers in social science use information asymmetry to represent the situation where one party in a transaction knows more compared to another, we would rather revise it as knowledge asymmetry in this paper for the purpose of being consistent with our technical framework of situation calculus and its extensions.

at the collective level and individual level respectively, which may be different among systems and agents.

Value is something that we think is important, and various types of values together with their orderings form a value system. By integrating the notion of value into our model, the result of performing opportunistic behavior is represented as the promotion of opportunistic individuals' value and the demotion of others' value. Furthermore, even though a value system is relatively stable within individuals, it may differ across different individuals and societies. For societies, each has its own value system as part of the social context and it serves as the basis for any judgment within the society. In this sense, some behaviors which are regarded as opportunistic in one society may not be considered as opportunistic in another society, if the two societies do not share the same value system. A similar idea, although more focusing on opportunistic propensity, can be found in [5]. Given the value system of the society, opportunistic behavior promotes the self-interest which is in opposition with others' value.

3   Technical Framework: Situation Calculus

Situation calculus provides a formal language for representing and reasoning about dynamical domains based on first-order logic. There are three types of sorts: actions that can be performed by agents, situations representing a history of action occurrences and objects for everything else. Situation S_0 represents the initial situation that no action can result in. The special predicate $do(a, s)$ denotes the unique situation that results from the performing of action a in situation s. The properties of situations are specified through relational and functional fluents taking a situation term as their last argument, which means their truth value may vary from situation to situation. The effects of actions on fluents are defined by successor state axioms. Also propositions P can be used instead of fluents, that is, their truth values are not dependent on the situation but consistent throughout all the situations.

With situation calculus, we can reason about how the world changes as the result of the available actions. A Basic Action Theory from Reiter [7] is defined as

$$D = \Sigma \cup D_{ap} \cup D_{ss} \cup D_{so} \cup D_{una}$$

Σ: the set of foundational axioms,

- $do(a_1, s_1) = do(a_2, s_2) \rightarrow a_1 = a_2 \wedge s_1 = s_2$;
- $(\forall Q)Q(S_0) \wedge (\forall s, a)[Q(s) \rightarrow Q(do(a, s))] \rightarrow (\forall s)Q(s)$;
- $s \sqsubset do(a, s') \equiv s \sqsubseteq s'$;
- $\neg s \sqsubset S_0$;

D_{ap}: the set of actions preconditions,

$$Poss(a(x), s) \equiv \pi_a(x, s)$$

D_{ss}: the set of successor state axioms,

$$F(do(a, s)) \equiv \gamma_F^+(a, s) \vee (F(s) \wedge \neg \gamma_F^-(a, s))$$

Here $\gamma_F^+(a, s)$ and $\gamma_F^-(a, s)$ are two formulas expressing the conditions for the fluent F becoming true and false, respectively; D_{so}: the sentences uniform in S_0 describing the initial situation; D_{una}: the unique name axioms for actions.

This is a brief overview of situation calculus, which is the technical preliminary of our formalization. However, this language can only provide information about the history of a situation and there is no way to represent the future of a situation. For example, propositions like "I sell the cup now" cannot be represented by situation calculus. Since this representation is of great importance to our formalization, we extend the situation to one-step further in the future. An extended situation is a pair (s, s') such that s is a situation and s' is the next situation of s, and *occur* is a relation between actions and situations. Here is the semantic of *occur*:

- $(s, s') \vDash occur(a, s)$ iff $s' = do(a, s)$. That is, $occur(a, s)$ holds if action a occurs in situation s.

After John McCarthy's introduction of this theory, people made extensions capable of representing knowledge, belief, intention and obligation in order to better reason about actions and their effects on the world [8–10]. We will introduce and adopt those extensions in the following sections as appropriate. Since in situation calculus the last argument is always a situation, we will follow this convention in this paper for any definition of fluents and predicates.

4   Formalizing Opportunism

4.1   Knowledge Asymmetry

We adopt the approach of Scherl to formalizing knowledge, which is to add an agents' possible-world model of knowledge to situation calculus [9]. To treat knowledge as a fluent, we have a binary relation $K(s', s)$, reading as situation s' is epistemically accessible from situation s. It is reflexive, transitive and symmetric.

Definition 4.1.1.

$$Know(i, \phi, s) \stackrel{def}{=} (\forall s')K_i(s', s) \rightarrow \phi[s']$$

This definition shows that an agent has knowledge about ϕ if and only if ϕ holds in all the epistemic possible situations of the agent. Then we can have the definition of knowledge asymmetry.

Definition 4.1.2.

$$KnowAsym(i, j, \phi, s) \stackrel{def}{=}$$
$$Know(i, \phi, s) \wedge \neg Know(j, \phi, s) \wedge Know(i, \neg Know(j, \phi, s), s)$$

KnowAsym is a fluent in situation s where agent i has knowledge about ϕ while agent j does not have and this is also known by agent i. It can be the other way around with i and j. But for defining opportunism, we only limit this definition to one case. Note that ϕ can represent any proposition in this definition.

4.2 Value Opposition

From the definition of opportunism, we know that agents have different evaluations on the same state transition. For agent i who performs opportunistic behavior, his value gets promoted, while the value of agent j gets demoted. We argue that this is because agents always make the evaluation from their perspective, which is part of their value system. This property of state transition is named value opposition in this study. In order to extend our technical framework with value theory, we define a symbol V to represent agents' value system and a binary relation $<$ over situations to represent agents' preference, where $s <_V s'$ denotes "s' is preferred to s based on value system V".

In situation calculus, situations can be described in terms of propositions P, which are structured with objects and their properties. For having preferences on situations, we argue that agents evaluate the truth value of specific propositions, which are called perspectives in this study, based on their value systems. For instance, the buyer tries to see if the cup has good quality or not in order to have a preference on the situations before and after the transaction. In order to specify agents' preference on situations, we first define a function $EvalRef$ that represents agents' perspective for evaluation:

Definition 4.2.1.
$$EvalRef : V \times S \times S \rightarrow P$$

It returns a proposition that an agent refers to for specifying his preference on two situations based on his value system. It is worth noting that in real life agents' specification of preferences on situations is based on a set of propositions 2^P rather than a single proposition. For instance, both whether the cup has good quality and appearance are important to the buyer. For simplicity, here we restrict the return value to only one proposition without loss of generality.

Once having this reference function, we can specify agents' preferences on situations, where V is restricted to perspective-based value:

$$s <_V s' \equiv Know(\neg p, s) \wedge Know(p, s') \text{ where } p = EvalRef(V, s, s')$$

$$s >_V s' \equiv Know(p, s) \wedge Know(\neg p, s') \text{ where } p = EvalRef(V, s, s')$$

It means that an agent's value gets promoted/demoted from s and s' when the truth value of the proposition p that the agent refers to based on his value system V changes. As for the example about selling the broken cup, the seller's value gets promoted when he knows that he has earned money from the transaction, whereas the buyer's value gets demoted when he knows that the cup is broken. Because of having different value systems, they refer to different propositions and thereby evaluate different propositions for specifying their preferences. Similar to knowledge asymmetry, we only limit the specification to one case for the truth value of p.

Definition 4.2.2.

$$ValueOppo(i, j, s, s') \stackrel{def}{=} s <_{V_i} s' \wedge s >_{V_j} s'$$

We define value opposition as a property of state transition where a state transition from s to s' can promote the value of agent i but demote the value of agent j. In other words, agent i has positive effects from the state transition, while agent j has negative effects. Again, we only limit the definition to one case for defining opportunism.

4.3 Intention

Opportunistic behavior is performed by intent rather than by accident. In order to suggest this aspect in our formal model, we adopt the logic of intention to do something for being something in our framework. The definition of $Intend$ is as below:

Definition 4.3.1.

$$Intend(i, a, \phi, s) \stackrel{def}{=} (\forall s') I_i(s', s) \rightarrow occur(a, s') \wedge \phi[s', do(a, s')]$$

$I(s', s)$ denotes the intentional accessibility relation of an agent, meaning that what is the case in situation s' is intended to have in situation s. $occur(a, s')$ is true when action a is performed in situation s', and ϕ is true in the state transition. An intention of agent i $Intend(i, a, \phi, s)$ holds if and only if both $occur(a, s')$ and $\phi[s', do(a, s')]$ hold in all intentional possible situations of agent i. Based on this definition of intention, we have two instances for value promotion $pro(j) = s' <_{V_j} do(a, s')$ and value demotion $de(j) = s' >_{V_j} do(a, s')$ by action a, which will be later used for providing the final definition and proving its properties

$$Intend(i, a, pro(j, v), s) \stackrel{def}{=} (\forall s') I_i(s', s) \rightarrow occur(a, s') \wedge s' <_{V_j} do(a, s')$$

$$Intend(i, a, de(j, v), s) \stackrel{def}{=} (\forall s') I_i(s', s) \rightarrow occur(a, s') \wedge s' >_{V_j} do(a, s')$$

$Intend(i, a, pro(j), s)$ denotes that agent i intends to promote the value of agent j by action a in situation s. Similar for $Intend(i, a, de(j), s)$. When $i = j$, agent i intends to promote or demote his own value by action a.

4.4 Opportunistic Behavior

The above definitions are pivotal ingredients that we need for having the formal model of opportunism: knowledge asymmetry as the precondition, value opposition as the effect, and intention as the mental state. Besides, based on the informal definition we gave in Sect. 2, there are two more aspects that should be suggested in the definition. Firstly, the Knowledge that the performer has while others do not have should be relevant to the state transition. Secondly, the performer is aware of value opposition for the state transition beforehand but still ignores it. Opportunism is defined as follows:

Definition 4.4.1. *Let D be a Situation Calculus BAT, K and I be the axioms for knowledge and intention representation in the Situation Calculus respectively, V be the value system of agents, EvalRef be the reference function representing the object for an agent's evaluation on situations, and $<_V$ be a preference ordering on situations. Then $(D \cup K \cup I, V, EvalRef, <_V)$ is a situation calculus BAT extended with knowledge, intention, value and preference. Within this system, we have*

$$Opportunism(i, j, a, s) \stackrel{def}{=} Poss(i, j, a, s) \wedge Intend(i, a, pro(i), s) \wedge \phi$$
$$where\ Poss(i, j, a, s) \equiv KnowAsym(i, j, \phi, s)$$
$$\phi = ValueOppo(i, j, s, do(a, s)).$$

This formula defines a predicate *Opportunism* where action a is performed by agent i to agent j in the situation s. In this concise formula, the precondition of action a is knowledge asymmetry about the state transition from s to $do(a, s)$, and action a is performed by intent and results in value opposition.

Another observation from the model is about the subjectivity of opportunism. We can see through the functional fluent Eval that agents always evaluate the situations and consequently the state transition from their own perspectives, which are part of their value systems. If the value systems upon which they have evaluation change to other ones, the property of value opposition may become false. Opportunism is presented as a "problem" in most research. However, the above formal model of opportunism implies that it depends on from which perspective, or more generally value systems, we evaluate the state transition. It is positive from the perspective of agent i, while it is negative from the perspective of agent j. It is not necessarily a good thing or a bad thing; it could be either. In reality and multi-agent systems, people usually take the established norms into consideration when they decide whether it should be prevented, and the result may be different from society to society and from system to system.

After having the formal model of opportunism, we show how the propositions we informally suggest in text at the beginning is captured by our formalization.

Proposition 4.4.1. *Given an opportunistic behavior performed by agent i to agent j, both agents evaluate the behavior from different perspectives.*

$$\vDash Opportunism(i, j, a, s) \rightarrow EvalRef(V_i, s, do(a, s)) \neq EvalRef(V_j, s, do(a, s))$$

Proof. If $Opportunism(i, j, a, s)$ holds, the property $ValueOppo(i, j, s, do(a, s))$ also holds. Following the definition of value opposition, we have

$$s <_{V_i} do(a, s) \wedge s >_{V_j} do(a, s).$$

The specification of $s <_{V_i} do(a, s)$ is

$$Know(i, \neg p, s) \wedge Know(i, p, do(a, s)) \text{ where } p = EvalRef(V_i, s, do(a, s)) \quad (1)$$

The specification of $s >_{V_j} do(a, s)$

$$Know(j, q, s) \wedge Know(j, \neg q, do(a, s)) \text{ where } q = EvalRef(V_j, s, do(a, s)) \quad (2)$$

Sentence (1) and (2) hold together. Since knowledge is consistent (i.e., the knowledge about something and the knowledge of its negation cannot exist at the same time), we have $p \neq q$, that is

$$EvalRef(V_i, s, do(a, s)) \neq EvalRef(V_j, s, do(a, s)). \qquad \square$$

Proposition 4.4.2. *Given an opportunistic behavior performed by agent i to agent j, agent i knows the performing of this behavior demotes agent j's value, but needs not intend to get this result, which is characterized by:*

$$\vDash Opportunism(i, j, a, s) \rightarrow Know(i, s >_{V_j} do(a, s), s)$$

$$\nvDash Opportunism(i, j, a, s) \rightarrow Intend(i, a, de(j, v), s)$$

Proof. The first one is already in the definition of opportunism, and we are going to prove the second one. Since the second one means that the implication does not hold in the model, what we need to do is to find a model where $Opportunism(i, j, a, s)$ is true whereas $Intend(i, a, de(j), s)$ is false. The model is given as follows.

Free riding is one of the classic models about opportunism, and it occurs when someone benefits from resources, goods, or services but does not pay for them, which results in either an under-provision of those goods or services, or in an overuse or degradation of a common property resource [25]. Suppose agent i is a free rider, then its behavior free riding satisfies the definition of opportunism. We have

$$Poss(i, others, freeride, s) \wedge Intend(i, freeride, pro(i), s) \wedge \phi$$
$$\text{where } Poss(i, others, freeride, s) \equiv KnowAsym(i, others, \phi, s)$$
$$\phi = ValueOppo(i, others, s, do(freeride, s)).$$

Then we have the two sentences below,

$$(\forall s')K_i(s', s) \rightarrow s' >_{V_{others}} do(freeride, s')$$

$$(\forall s')I_i(s', s) \rightarrow occur(freeride, s') \wedge s' <_{V_i} do(freeride, s')$$

which mean that agent i knows his behavior will demote the value of others, and it is his intention to promote his value by free riding.

However, the following sentence, which means that it is agent i's intention to demote the value of others, does not hold in our model,

$$(\forall s')I_i(s', s) \rightarrow occur(freeride, s') \wedge s' >_{V_{others}} do(freeride, s')$$

It is firstly because, in our formalization, we define K-relation and I-relation as two distinct types of accessibility relations such that something holds in the possible situations of knowledge does not mean that it holds in the possible situations of intention as well. Secondly, at the empirical level, agent i does not intend to reduce others' share of public goods. Therefore, $Intend(i, freeride, de(others), s)$ does not hold in our model.   □

4.5   Opportunistic Behavior for Multiple Actions

In the previous section, we only consider one single action as opportunistic behavior. But in real life it is common that opportunistic behavior consists of multiple actions. For instance, unlike the simple selling example at the beginning of this paper, commerce transactions between businesses usually have a couple of actions, each of which ends up with a status. In this context, the sequence of actions is opportunistic behavior instead of any single action within. Of course, a sequence of actions can be seen as one action if we only look at the precondition of the first action and the effect of the last action, but we may also investigate what properties we can derive from opportunistic behavior for multiple actions.

In situation calculus, a binary function $do(a, s)$ is used to denote the situation resulting from performing action a in situation s, so for a finite sequence of actions $[a_1, \ldots, a_n]$, the situation resulting from performing the sequence of actions in situation s is denoted as $do(a_n, do(a_{n-1}, \ldots, do(a_1, s)))$. Therefore, based on Definition 4.1.1, the formal model of opportunistic behavior for multiple actions is given as below:

Definition 4.5.1. *Let D be a Situation Calculus BAT, K and I be the axioms for knowledge and intention representation in the Situation Calculus respectively, V be the value system of agents, EvalRef be the reference function representing the object for an agent's evaluation on situations, and $<_V$ be a preference ordering on situations. Then $(D \cup K \cup I, V, EvalRef, <_V)$ is a situation calculus BAT extended with knowledge, intention, value and preference. Within this system, we have*

$$Opportunism(i, j, [a_1, \ldots, a_n], s_1) \overset{def}{=}$$
$$\bigwedge_{1 \le k \le n} Poss(i, j, a_k, s_k) \land Intend(i, a_k, pro(i), s_k) \land \phi$$
$$where\ Poss(i, j, a_k, s_k) \equiv KnowAsym(i, j, \phi, s_k)$$
$$\phi = ValueOppo(i, j, s_1, do(a_n, do(a_{n-1}, \ldots, do(a_1, s_1))))$$
$$s_k = do(a_{k-1}, \ldots, do(a_1, s_1))(1 < k \le n)$$

Because each action in the sequence must be possible to be performed and it is the property of intention to be persistent [27], knowledge asymmetry and intention is true in s_k for $1 \le k \le n$. Value opposition is the property of the state transition by the sequence of actions. A finite sequence of actions $[a_1, \ldots, a_n]$, which is performed by agent i to agent j in situation s_1, is opportunistic behavior if and only if each action is possible with the intention to promote agent i's value and the whole sequence results in value opposition for agent i and j. This definition captures some interesting properties, which cannot be derived from Definition 4.4.1.

Proposition 4.5.1. *Given a finite sequence of actions $[a_1, \ldots, a_n]$ as opportunistic behavior, we can prove that*

$$\models Opportunism(i, j, [a_1, \ldots, a_n], s_1) \rightarrow$$
$$KnowAsym(i, j, \phi, s_k) \equiv KnowAsym(i, j, \phi, do(a_k, s_k))(1 < k < n)$$

Proof. Each action in the sequence is possible to be performed and also

$$Poss(i, j, a_k, s_k) \equiv KnowAsym(i, j, \phi, s_k)(1 \leq k \leq n)$$

$$s_k = do(a_{k-1}, \ldots, do(a_1, s_1))(1 < k \leq n)$$

Combining these two formulas, we can easily get

$$KnowAsym(i, j, \phi, s_k) \equiv KnowAsym(i, j, \phi, do(a_k, s_k))(1 \leq k < n). \qquad \square$$

This proposition shows that, when opportunistic behavior consists of a sequence of actions, property knowledge asymmetry is preserved throughout the whole sequence.

Proposition 4.5.2. *Given a finite sequence of actions $[a_1, \ldots, a_n]$ as opportunistic behavior, we can prove action a_i needs not be opportunistic, which is characterized by*

$$\nvDash Opportunism(i, j, [a_1, \ldots, a_n], s_1)(n > 1) \rightarrow$$
$$Opportunism(i, j, a_k, s_k)(1 \leq k \leq n)$$

Proof. In order to prove this proposition, we are going to find a counterexample of opportunistic behavior which satisfies $n > 1$ but each action does not satisfy all the properties of opportunism.

Freeriding is still a nice model to prove this property. Since freeriding is one form of opportunistic behavior, $Opportunism(i, others, freeride, s_1)$ is true in our model. Now we are going to split it into a sequence of actions $[a_1, \ldots, a_n]$ and suppose a free rider exist in a society with a large population and benefits from the public goods without paying. Since the amount that the free rider is supposed to pay is shared by a large population, other agents do not notice (or even not care about) the small change thus not getting their value demoted for little amount of freeriding. That is, for action a_k,

$$Know(others, q, do(a_k, s_k)) \rightarrow \neg(s_k >_{V_j} do(a_k, s_k))$$

holds, where $q = EvalRef(V_{others}, s, do(a_k, s_k))$. Therefore, it is not true that

$$Opportunism(i, j, a_k, s_k)(1 \leq k \leq n). \qquad \square$$

However, once the amount that the free rider is supposed to pay accumulates to be large enough for getting other agents' value demoted (the whole sequence of actions is considered) will it be regarded as opportunistic behavior. By theoretical comparison, this example is quite similar to Sorites paradox, where grains are individually removed from a heap of sands and the heap stops being a heap when the process is repeated for enough times [11]. So it is also interesting to think about when the behavior starts to be regarded as opportunistic.

In the example above, the fact associated with agents' preference is ignored for its small change. It is also possible that the information associated with an agent's preference is blocked. In this case, he cannot specify his preference on the situations and consequently cannot evaluate the actions. Only when he receives the specific information and compares his current situation with previous situations can the sequence of actions be considered as opportunistic behavior.

4.6   Opportunistic Behavior with Social Context

In the previous sections, we made an assumption for the sake of simplicity that there is no legal or moral evaluation being made or implied to opportunistic behavior such that it is not necessary good or bad. However, agents in MAS are residing in a social context which provides obligations, permissions and other types of norms for guiding agents' behaviors. The setting of those norms can also reflect the value system of a society. To have a formal model of opportunism with social context, we can of course replace the agent j in our previous models with a society (in this way, we see the whole society as an agent) and get similar properties as in last two sections, but now we are more interested in putting opportunism in a deontic-based social context to see how it relates to social norms. Thus, in this section, we are going to put opportunistic behavior into a social context with norms and propose a formal model of opportunism from this perspective.

For defining opportunistic behavior with social context, we adopt the definition of knowledge asymmetry and intention in previous sections but redefine value opposition. Firstly, we have three normative statuses, which are similar to deontic logic.

- it is obligatory that (OB)
- it is permissible that (PE)
- it is forbidden that (FO)

Secondly, we define the above deontic notions for specifying the normative propositions Π.

Definition 4.6.1.

$$OB(i, a, s) \stackrel{def}{=} (\forall s')R_i(s', s) \rightarrow occur(a, s')$$

$$PE(i, a, s) \stackrel{def}{=} (\exists s')R_i(s', s) \wedge occur(a, s')$$

$$FO(i, a, s) \stackrel{def}{=} (\forall s')R_i(s', s) \rightarrow \neg occur(a, s')$$

where $R_i(s', s)$ denotes the deontic accessibility relation of agent i, meaning that what is the case in situation s' is ideal for situation s, and $occur(a, s')$ is true when action a is performed in situation s'. R-relation is serial, which means for all situations s there is at least one possible situation s' such that $R_i(s', s)$ holds. This property of R-relation ensures $OB(i, a, s) \rightarrow PE(i, a, s)$ to be hold, which is also consistent with our intuition. Each modality can be taken as a basic to define the other two modalities.

We then specify the social preference on situations, where V is restricted to deontic-based social value.

$$s <_{V_A} s' \equiv (\exists a, i)s' = do(a, s) \wedge OB(i, a, s)$$

$$s >_{V_A} s' \equiv (\exists a, i)s' = do(a, s) \wedge FO(i, a, s)$$

Here symbol A represents the whole society, which is a set of agents. The first one means that the social value gets promoted if there exists an action whose performing complies with the social norm, while the second one means that the social value gets demoted if there exists an action whose performing violates the social norm.

Together with the specification of agents' preferences on situations, we have the definition of value opposition between agents and the whole society.

Definition 4.6.2.

$$ValueOppo(i, A, s, s') \stackrel{def}{=} s <_{V_i} s' \wedge s >_{V_A} s'$$

For the state transition from s to s', the value of agent i gets promoted whereas the social value gets demoted. We only limit the definition to one case excluding the other way around for defining opportunism.

Therefore, similar to Definition 4.4.1, we have the definition of opportunistic behavior with social context.

Definition 4.6.3. *Let D be a Situation Calculus BAT, K and I be the axioms for knowledge and intention representation in the Situation Calculus respectively, V be the value system of agents, $EvalRef$ be the reference function representing the object for an agent's evaluation on situations, Π be a finite set of normative propositions, and $<_V$ be a preference ordering on situations. Then $(D \cup K \cup I, V, EvalRef, \Pi, <_V)$ is a situation calculus BAT extended with knowledge, intention, value, norms and preference. Within this system, we have*

$$Opportunism(i, A, a, s) \stackrel{def}{=} Poss(i, A, a, s)Intend(i, a, pro(i), s) \wedge \phi$$

$$where\ Poss(i, A, a, s) \equiv KnowAsym(i, A, \phi, s)$$

$$\phi = ValueOppo(i, A, s, do(a, s)).$$

Action a performed by agent i is regarded as opportunistic behavior if and only if it is performed with the asymmetric knowledge ϕ about the state transition from s to $do(a, s)$ and the intention of self-interest, and results in value opposition against the society A where he is staying.

The definition of opportunistic behavior with social context shows that, given the value system of a society, opportunistic behavior is considered to be bad since its performing results in demoting the social value. Further, it implies the moral dilemma concerning the conflict between desire and obligation. More precisely, an agent has the desire "to do what one wants", while the social context where the agent is residing gives the obligation "to do what one ought to do". Opportunistic agents follow their desire but ignore the obligation. Hence, it should be prohibited by laws or social norms from the perspective of the whole society.

Since we assume a social context with norms in this section, it is worth investigating the relation between deontic notions and mental states. Our formalization governs Proposition 4.6.1 regarding opportunistic agents having knowledge about the relevant norms, and Proposition 4.6.2 and Proposition 4.6.3 about the intention of opportunistic behavior not being derived from the obligation.

Proposition 4.6.1. *Let action a be opportunistic behavior performed by agent i within society A in situation s, then for the social norm associated with action a $FO(i, a, s) \in \Pi$ we have*

$$\vDash Opportunism(i, A, a, s) \rightarrow Know(i, FO(i, a, s), s)$$

Proof. Since $Opportunism(i, A, a, s)$ holds, by Definition 4.6.3, agent i must have knowledge about the effect of performing action a, that is, $Know(i, \phi, s)$ holds, where ϕ represents value opposition. By Definition 4.6.2, $\phi = s <_{V_i} do(a, s) \wedge s >_{V_A} do(a, s)$. Therefore, $Know(i, s >_{V_A} do(a, s), s)$ holds. Because V is restricted to deontic-based social value in our model, $s >_{V_A} do(a, s) \equiv FO(a, s)$ holds, thereby $Know(i, FO(i, a, s), s)$ holds as well. ☐

Agents have the knowledge about the relevant norms in the society and decide whether and which to comply with based on their own analysis. Typically, opportunistic agents behave in their interest, regardless of the social norms they are supposed to follow.

Moreover, intention might be derived from obligation (I ought to go to work this morning, so I intend to go to work this morning), or might just come from agents' own desire (I feel thirsty, so I intend to get some water). In a given situation, agents intend to perform opportunistic behavior, which is motivated by self-interest. In order to prove this property rigorously, we should first prove the disobedience of opportunistic behavior.

Proposition 4.6.2. *Let action a be opportunistic behavior performed by agent i within society A in situation s, and V_i be agent i's value system, we can prove*

$$\vDash Opportunism(i, A, a, s) \rightarrow (V_i \neq Obedience)$$

Proof. By contradiction, we assume that $V_i = Obedience$. Because agent i obeys to the social norm in order to promote his value, action a should not be forbidden by the society, that is, $FO(a, s)$ does not hold. Consequently, $s >_{V_A} do(a, s)$ and $Opportunism(i, A, a, s)$ do not hold, either. Therefore, $V_i = Obedience$ is false for opportunistic behavior. ☐

Using Proposition 4.6.2, we are going to prove it does not hold in opportunistic behavior that the intention is derived from the obligation.

Proposition 4.6.3. *Let action a be opportunistic behavior performed by agent i within society A in situation s, then for the social norm associated with action a $OB(i, a, s) \in \Pi$, we can prove*

$$\vDash Opportunism(i, A, a, s) \rightarrow \neg(OB(i, a, s) \rightarrow Intend(i, a, pro(i), s))$$

Proof. We can prove this proposition by contradiction. Suppose action a is opportunistic behavior and sentence $OB(i, a, s) \rightarrow Intend(i, a, pro(i), s)$ holds in our model, which means the intended situations of agent i are the subset of ideal situations, formalized as $(\forall s')I_i(s', s) \rightarrow R_i(s', s)$, where s' in $I_i(s', s)$ is restricted to the situation that satisfies $pro(i)$. Therefore, agent i intends to promote his own value and the social value by action a. Of course, when agent i's value is obedience, both agent i's value and the social value are promoted. But we have already proved in Proposition 4.6.2 that this possibility does not exist. So our assumption at the beginning is wrong. Therefore, $OB(i, a, s) \rightarrow Intend(i, a, pro(i), s)$ does not hold in our model.   □

5   Example: Selling a Broken Cup

Recall the example that we used to introduce opportunism at the beginning of the paper. The scenario is simple but enough to illustrate our formal model of opportunism. We label the seller and the buyer as s and b, who can be in one of the situations: S_0 (the initial situation, before the transaction) and $do(a, S_0)$ (after the transaction). The seller can either sell the cup ($a = sell(x)$) or keep it. If the seller performs the action $sell(x)$ in S_0, then situation will go to $do(sell(x), S_0)$.

In situation S_0, the asymmetric knowledge owned by the seller but not the buyer is not only about the broken cup, but also the state transition: once the transaction finishes, the situation will go from S_0 to $do(sell(x), S_0)$, which gets the value of the seller promoted whereas the value of the buyer demoted. That is, the precondition $KnowAsym(s, b, \phi, S_0)$ holds. Now consider the value for both parties. Apparently both parties go for economic value. However, they have different and contradictory perspectives about the economic value. What the seller looks at is how much money he earns from the transaction. When the seller knows the broken cup has already been sold, his value gets promoted ($S_0 <_{V_s} do(sell(x), S_0)$ holds). Conversely, what the buyer looks at is whether the cup has good quality or not. So when the buyer knows the cup is broken, his value gets demoted ($S_0 >_{V_b} do(sell(x), S_0)$ holds). The above two sentences ensure $ValueOppo(s, b, S_0, do(sell(x), S_0))$. Further, since it is the seller's intention to sell the broken cup to the buyer for promoting his value, sentence $Intend(s, sell(x), pro(s), S_0)$ also holds. With the above formalization, we have the formula of opportunistic behavior for this example $Opportunism(s, b, sell(x), S_0)$.

From the above illustration, we can think of two situations that are worth considering. Firstly, if the buyer buys the cup only for decoration without using it, he will never know the cup is broken or even not care about it. In other words, the perspective of the seller getting the cup sold and the perspective of the buyer that the cup is good for decoration are not contradictory. That is, the buyer's perspective is revised to $EvalRef(V_b, S_0, do(sell(x), S_0)) = appearance$. Because the cup has good appearance, sentence $S_0 >_{V_b} do(sell(x), S_0)$ does not hold. In this case, the seller's behavior may not be opportunistic from the perspective of the buyer, if the social norms are not taken into account. The subjectivity of opportunism is reflected by the different judgement on the same action. Secondly, if there is nothing the seller can do except sell the broken cup when being in state S_0, it will be regarded as opportunistic behavior with the nature of self-defense based on our first formal definition. It is because there is no moral or legal evaluation in this definition thus no matter whether the behavior is good or bad. In this sense, we can assume social context and analyze it with Definition 4.6.3. Suppose self-defense behavior is allowed by the society ($PE(i, a, S_0)$). Then $S_0 >_{V_A} do(sell(x), S_0)$ does not hold. Therefore, it is not opportunistic behavior from the perspective of the society. In our example, however, the options available to the buyer in state S_0 are $\{sell, keep\}$, which means selling the broken cup is not the only action that he can perform.

Further, with the help of our model, we can gain practical insights into constraint mechanism of opportunism. In our case, one important reason why the seller's behavior is seen as opportunistic is that the seller and the buyer evaluate the state transition from two contradictory perspectives based on their value systems. In other words, even though they both go for economic value, they look at different things for evaluation. When applying this approach in collaborative relationship, it is much easier to understand how the relationship ends in defection. Therefore, one deterrence mechanism for partner opportunism is to avoid having contrasted value systems in the relationship. As for the precondition of opportunism, even though it is difficult to prevent knowledge asymmetry in business transactions, we still need to think about how much information we can provide to our partners and how they are going to use the information.

6   Discussion

As it is the first step of our work, we try to propose a simple but elegant model of opportunism for different context settings by making restricted assumptions. But it also means that the model might not manage to capture every possible scenarios. For instance, in Sect. 4.2 we only talk about the interaction between two agents and investigate the evaluation on the state transition from the perspectives of the two agents who are involved in the transaction. But actually such evaluation can also be done by others. Assume that a person sees the transaction and his value system is incompatible with agent i's. He may get angry with the seller even though he is not involved. In this sense, the behavior that is performed by agent i is considered to be opportunistic from the perspective of the third agent. Further, our models only consider intentional actions. However, opportunistic behavior can also be about intentional inaction, which should really have been taken for obligation or responsibility, such as hiding important information. In this case, the social value gets demoted for agent i's not performing an obligatory action instead of performing a forbidden action. Of course, our models can capture this scenario in a way that deliberate inaction can be seen as an action. Interesting insights can be gained from further study on this part.

We also propose that the asymmetric knowledge obtained by opportunistic agents is value opposition about the state transition, which not the same as our intuition. The reason can be shown by the example in Sect. 5. Intuitively the asymmetric knowledge that the seller has is about the broken cup. Now we assume that both the seller and the buyer know the cup is broken and the seller sells it with a high price. Once the buyer knows that the broken cup is not worth that price, his value will get demoted. From that we can see it does not matter whether the fact about the broken cup is only known by one party beforehand, but whether value opposition about the transaction is only known by one party beforehand. In other words, the asymmetric knowledge is not about the objective fact, but about agents' evaluation on the state transition.

7   Related Work

Opportunism is not a new topic in social science. Since it was released by economist Williamson, scholars have studied the typical social behavior of economic players from various perspectives i.e. transaction cost economics [13], resource-based view [14], game theory [4], agency theory [15] and strategic management [16]. Even though they are indeed all worthwhile, it is difficult to directly apply their conclusions to MAS for improving the system's behavior because most of them are informal, which makes reasoning about this behavior in MAS impossible, and also not commonly accepted even in their own area.

In the field of artificial intelligence, there is a tradition to devise intelligent artifacts and construct intelligent system using symbolic representation of all factors involved [17]. Especially for mathematical logic, it is a greatly important approach to this field due to its highly abstract representation and reasoning about social reality. Therefore, a lot of work on logic formalism has been designed for representing and reasoning about dynamical domains such as situation calculus [18], event calculus [19] and fluent calculus [20]. We chose to use situation calculus as our basic framework because it has been well developed and extended with knowledge [9], belief [8] and other model semantics. In [9], an epistemic fluent $Know(P, s)$ is proposed by adapting the standard possible-world model of knowledge to situation calculus. We use this approach to define knowledge asymmetry where agents possess different amount of knowledge.

We integrated the notion of value into situation calculus to represent agents' preference on situations. However, in logical formalization, people usually use goals rather than value (e.g. [12,21]) for the same purpose. Only some works in the area of argumentation reason about agents' preferences and decision making by value (e.g. [22,23]). Even though both goals and values can be used to reason about agents' preferences about situations, they have different features. Goals are concrete and should be specified with time, place and objects. For example, to earn 1000 euros next month is a goal. If one agent's goal is achieved in one situation, then he has high evaluation on that situation. Value is described by Schwartz as trans-situational [24], which means that value is relatively stable and not limited to be applied in a specific situation. For instance, if honesty is a value of somebody, he will be honest for a long period of time. Since state transition results from the performing of actions, we can evaluate actions by whether our value is promoted or demoted in the state transition, as what we do in this study. For representing agents' evaluation on situations, Keeney and Raiffa proposed Multi-Attribute Utility Theory (MAUT) in which situations are described in terms of a set of attributes and the utilities of the situations are calculated by the sum of the scores on each attribute based on agents' value system [26]. In this study, we use a similar approach in which situations are represented through propositions and agents focus on a specific proposition based on their value systems to evaluate a state transition. Apparently, different agents may focus on different propositions thus having different evaluations on the same state transition.

8   Conclusions

Agents situated with information asymmetry might perform opportunistic behavior to others in their interest. Numerous works about such a social behavior have been seen in social science due to its negative effect on the relationship. However, most conclusions are based on a given form of opportunism, making it hard to build a fundamental theory that can be applied in any contexts. This study took the initiative to propose a formal model of opportunism based on the extended informal definition from Williamson. The modeling work was done based on situation calculus integrating the notion of value. We first have a preliminary model that only considers a single action between two agents, and then extend it for multiple actions with social context. Each model captures interesting properties that useful for our future research.

It is important to keep in mind that our aim is not to indicate where opportunistic behavior comes from through the model we propose, as before coming to this part we should have a thorough understanding of the nature of opportunism. Therefore, the main strength of this study is defining such a behavior from our specific perspective in a formal way, so as to represent the elements in the definition and how they relate to each other, towards building a formal system of opportunism. Only when we have a formal system can we perform further investigation on its emergence and constraint mechanism.

Further study can be carried out on the state transition within our formal system. For having a formal definition of opportunism, we defined a set of fluents that characterize the situations before and after the action is performed. However, we still have no idea how the truth value of each fluent is changed in the successor situation. For instance, fluent knowledge asymmetry is true in the initial situation for being a precondition of opportunism, but with our formalization we cannot prove its truth value after the action is performed. This problem can be solved by having successor state axioms for each fluent we define, which is important to representing and reasoning about the dynamics of our formal system. Of course, it is not necessary to do that for just having a formal definition. Another avenue would be to investigate how opportunism emerges based on the definition of opportunism. As we mentioned in our example, agents are not able to perform opportunistic behavior if the precondition knowledge asymmetry fails. However, it is common that agents stay in different positions with a different amount of information, so knowledge asymmetry is unavoidable. Therefore, we need to think about how much information and what kinds of information we can share with our partners. Moreover, agents' having different perspectives on the same value is the reason to value opposition of a state transition. So it is natural to think about how agents evaluate a situation from their perspectives and how the perspectives relate to their value systems for the study of opportunism emergence. Considerable insights can be achieved from the investigation about the compatibility of different value systems and the co-evolution of agents' value system with social context or environmental changes.

Acknowledgments. The research is supported by China Scholarship Council. We would like to thank anonymous reviewers, Marlo Souza, Hein Duijf and other colleagues for their helpful comments.

References

1. Williamson, O.E.: Markets and Hierarchies, Analysis and Antitrust Implications: A Study in the Economics of Internal Organization. Free Press, New York (1975)
2. Wooldridge, M.: An Introduction to Multiagent Systems. Wiley, Chichester (2009)
3. Das, T.K., Rahman, N.: Determinants of partner opportunism in strategic alliances: a conceptual framework. J. Bus. Psychol. **25**(1), 55–74 (2010)
4. Cabon-Dhersin, M.-L., Ramani, S.V.: Opportunism, trust and coopera-tion a game theoretic approach with heterogeneous agents. Rationality Soc. **19**(2), 203–228 (2007)
5. Chen, C.C., Peng, M.W., Saparito, P.A.: Individualism, collectivism, and opportunism: a cultural perspective on transaction cost economics. J. Manag. **28**(4), 567–583 (2002)
6. McCarthy, J.: Situations, actions, and causal laws. In: Minsky, M. (ed.) Semantic Information Processing, pp. 410–417. MIT Press, Cambridge (1968)
7. Reiter, R.: Knowledge in Action: Logical Foundations for Specifying and Implementing Dynamical Systems, vol. 16. MIT press, Cambridge (2001)
8. Shapiro, S., Pagnucco, M., Lesprance, Y., Levesque, H.J.: Iterated belief change in the situation calculus. Artif. Intell. **175**(1), 165–192 (2011)
9. Scherl, R.B., Levesque, H.J.: Knowledge, action, and the frame problem. Artif. Intell. **144**(1), 1–39 (2003)
10. Demolombe, R., Parra, P.P.: Integrating state constraints and obligations in situation calculus. Inteligencia Artificial, Revista Iberoamericana de Inteligencia Artificial **13**(41), 54–63 (2009)

11. Hyde, D.: Sorites paradox. In: Zalta, E.N.(ed.).: The Stanford Encyclopedia of Philosophy (Winter 2014 Edition). http://plato.stanford.edu/archives/win2014/entries/sorites-paradox/
12. Cohen, P.R., Levesque, H.J.: Intention is choice with commitment. Artif. Intell. **42**(2), 213–261 (1990)
13. Williamson, O.E.: The Economic Institutions of Capitalism. Free Press, New York (1985)
14. Conner, K.R., Prahalad, C.K.: A resource-based theory of the firm: knowledge versus opportunism. Organ. Sci. **7**(5), 477–501 (1996)
15. Jiraporn, P., et al.: Is earnings management opportunistic or beneficial? an agency theory perspective. Int. Rev. Financ. Anal. **17**(3), 622–634 (2008)
16. Yaqub, M.Z.: Antecedents, consequences and control of opportunistic behavior in strategic networks. J. Bus. Econ. Res. (JBER) **7**(2), 15–31 (2009)
17. Segerberg, K., Meyer, J.-J., Kracht, M.: The logic of action. In: Zalta, E.N.(ed.).: The Stan-ford Encyclopedia of Philosophy (Winter 2013 Edition). http://plato.stanford.edu/archives/win2013/entries/logic-action/
18. McCarthy, J., Hayes, P.: Some philosophical problems from the standpoint of artificial intelligence. Stanford University, USA (1968)
19. Kowalski, R., Sergot, M.: A logic-based calculus of events. In: Schmidt, J.W., Thanos, C. (eds.) Foundations of Knowledge Base Management: Contributions from Logic, Databases, and Artificial Intelligence Applications. Topics in Information Systems, pp. 23–55. Springer, Heidelberg (1989)
20. Thielscher, M.: Introduction to the fluent calculus. Electron. Trans. Artif. Intell. **2**(3–4), 179–192 (1998). http://www.etaij.org
21. Rao, A.S., Georgeff, M.P.: Modeling rational agents within a BDI-architecture. In: KR 1991, pp. 473–484 (1991)
22. Bench-Capon, T., Atkinson, K., McBurney, P.: Using argumentation to model agent decision making in economic experiments. Auton. Agent. Multi Agent Syst. **25**(1), 183–208 (2012)
23. van der Weide, T.L.: Arguing to motivate decisions. SIKS Dissertation Series 2011 (2011)
24. Schwartz, S.H.: Universals in the content and structure of values: theoretical advances and empirical tests in 20 countries. In: Advances in experimental social psychology, vol. 25, pp. 1–65 (1992)
25. Baumol, W.J.: Welfare Economics and the Theory of the State. Harvard University Press, Cambridge (1967)
26. Keeney, R.L., Raiffa, H.: Decisions with Multiple Objectives: Preferences and Value Trade-offs. Cambridge University Press, Cambridge (1993)
27. Bratman, M.: Intention, Plans, and Practical Reason. Harvard University Press, Cambridge (1987)

Programming JADE and Jason Agents Based on Social Relationships Using a Uniform Approach

Matteo Baldoni[✉], Cristina Baroglio, and Federico Capuzzimati

Dipartimento di Informatica, Università degli Studi di Torino,
c.so Svizzera 185, 10149 Torino, Italy
{matteo.baldoni,cristina.baroglio,federico.capuzzimati}@unito.it

Abstract. Interaction is an essential feature in multiagent systems. Design primitives are needed to explicitly model desired patterns. This work presents 2COMM as a framework for defining social relations among parties, represented by social commitments. Starting from the definition of interaction protocols, 2COMM allows to decouple interaction design from agent design. Currently, adapters were developed for allowing the use of 2COMM with the JADE and the JaCaMo platforms. We show how agents for the two platforms can be implemented by relying on a common programming schema.

Keywords: Social commitments · Agents and Artifacts · Agent-oriented software engineering

1 Introduction and Motivation

Multiagent Systems (MAS) are a preferred choice for building complex systems where the autonomy of each component is a major requirement. Agent-oriented software engineers can choose from a substantial number of agent platforms [5,11,16,18]. Tools like JADE [8], TuCSoN [20], DESIRE [12], JaCaMo [10], all provide coordination mechanisms and communication infrastructures [11] but, in our opinion, they lack of abstractions that allow a clear and explicit modeling of interaction. Basically, the way in which agents interact is spread across and "hard-coded" into agent implementations. This choice overly ties agent implementations with a negative impact on software reuse. A clear separation of the agents from the specification of their coordination would bring advantages both on the design and on the implementation of MAS by bringing in a greater decoupling.

To this purpose, we propose to explicitly represent agent coordination patterns in terms of normatively defined *social relationships*, and to ground this normative characterization on *commitments* [23] and on commitment-based interaction protocols [25]. Practically, we exploit 2COMM [1], a tool that allows building artifacts that incorporate commitment-based protocols, in a way that is not bounded to an agent platform, and two connectors that allow using 2COMM for making JADE agents and JaCaMo agents interact.

© Springer International Publishing Switzerland 2015
F. Koch et al. (Eds.): CARE-MFSC 2015, CCIS 541, pp. 167–184, 2015.
DOI: 10.1007/978-3-319-24804-2_11

Relying on artifacts has the advantage of transforming social relationships and coordination patterns into *resources* and this allows agents to dynamically recognize, accept, manipulate, reason on them, and decide whether to conform to them (a basis for coordination [15]). In order to reify the social relationships we rely on the Agents and Artifacts meta-model (A&A) [19,24], which provides abstractions for environments and artifacts, that can be acted upon, observed, perceived, notified, and so on. 2COMM adopts the abstraction of artifact to construct communication protocols that realize a form of mediated, programmable communication, and in particular commitment protocols to establish an interaction social state agents can use to take decisions about their behaviour. Through 2COMM protocol artifacts, social relationships can be examined by the agents, as advised in [14], used (which entails that agents accept the corresponding regulations), constructed, e.g., by negotiation, specialized, composed, and so forth. Finally, 2COMM artifacts enable the implementation of monitoring functionalities for verifying that the on-going interactions respect the commitments and for detecting violations and violators.

This work also shows how starting from interactions in building a system can be useful when programming socially-responsive agents. A clear specification of the commitments that an agent has to handle constitutes an outline, for agent implementation, that developers can follow.

Summarizing, this work proposes to introduce in MAS an explicit notion of social relationship, based on that of commitment (Sect. 2). The framework 2COMM (Sect. 3), an extension of JaCaMo, realizes the proposal. We explain programming schemas for JADE and Jason agents. We show the impact of the proposal on programming by means of an example (Sect. 4) based on the FIPA Contract Net Protocol (CNP). The example shows (1) practical advantages in terms of better code organization and easier coding of agents interaction, and (2) how agent implementation is lead by the interaction pattern, providing a cross-platform programming pattern.

2   Modeling Social Relationships

We propose to explicitly represent social relationships among the agents. By social relationships we mean normatively defined relationships, between two or more agents, resulting from the enactment of *roles*, and subject to *social control*. Thus, we encode social relationships as commitments. A commitment [22] is represented with the notation $C(x, y, r, p)$, capturing that the agent x commits to the agent y to bring about the consequent condition p when the antecedent condition r holds. Antecedent and consequent conditions generally are conjunctions or disjunctions of events and commitments. When r equals $\top$, we use the short notation $C(x, y, p)$ and the commitment is said to be *active*. Commitments have a *regulative* nature, in that debtors are expected to behave so as to satisfy the engagements they have taken. This practically means that an agent is expected to behave so as to achieve the consequent conditions of the active commitments of which it is the debtor.

We envisage both agents and social relationships as first-class entities that interact in a bi-directional manner. Social relationships are created by the execution of *interaction protocols* and provide expectations on the agents' behaviour. It is, therefore, necessary to provide the agents the means to create, to manipulate, to observe, to monitor, to reason, and to deliberate on social relationships. We do so by exploiting *properly defined artifacts*, that reify both *interaction protocols*, defined in terms of social relationships, and the sets of *social relationships*, that are created during the protocols execution, available to agents as resources.

An artifact (A&A meta-model [19,24]) is a computational, programmable system resource, that can be manipulated by agents, residing at the same abstraction level of the agent abstraction class. For their very nature, artifacts can encode a mediated, programmable and observable means of communication and coordination between agents. We interpret the fact that an agent uses an artifact as its explicit acceptance, of the implications of the interaction protocol that the artifact reifies. This allows the interacting parties to perform *practical reasoning*, based on expectations: a debtors of a commitment is expected to behave so as to satisfy the commitment consequent conditions; otherwise, a violation is raised.

A commitment-based protocol consists of a set of actions, whose semantics is shared, and agreed upon, by all of the participants to the interaction [13,25]. The semantics of the social actions is given in terms of commitment operations (as usual for commitments, *create, cancel, release, discharge, assign,* and *delegate*). The execution of commitment operations modifies the *social state* of the system, which is shared by the interacting agents. As in [22], we postulate that discharge is performed concurrently with the actions that lead to the given condition being satisfied and causes the commitment to not hold. Delegate and assign transfer commitments respectively to a different debtor and to a different creditor [13,22,25]. Commitment-based protocols provide a means of coordination, based on the *notification* of social events, e.g. the creation of a commitment. Agents use artifacts to coordinate and interact in a way that depends on the roles they play and on their objectives.

From an organizational perspective, a protocol is structured into a set of *roles*. We assume that roles cannot live autonomously: they exist in the system in view of the interaction. We follow the ontological model for roles proposed in [9], and brought inside the object-oriented paradigm in [6,7], which is characterized by three aspects: (1) *Foundation*: a role must always be associated with the institution it belongs to and with its player; (2) *Definitional dependence*: the definition of the role must be given inside the definition of the institution it belongs to; (3) *Institutional empowerment*: the actions defined for the role in the definition of the institution have access to the state of the institution and of the other roles, thus, they are called powers; instead, the actions that a player must offer for playing a role are called requirements. The agents that will be the *role players* become able to perform protocol actions, that are *powers* offered by a specific role and whose execution affect the social state. On the other hand, they need to satisfy the related *requirements*: specifically, in order to play a role an agent needs to have the capabilities of satisfying the related commitments – capabilities which can be internal of the agent or supplied as powers as well.

3   2COMM: A Commitment-Based Infrastructure for Social Relationships

We have claimed that an agent-based framework should satisfy two require-ments: (1) Explicit representation of the social relationship; (2) Social relation-ships should be first-class objects, which can be used for programming the agent behavior. 2COMM fulfills both requirements. Thanks to the *social relationship* abstraction, 2COMM enables an approach to agent programming that is not coupled to the chosen agent platform.

Currently, 2COMM supports social relationship-based agent programming for *JADE* and *JaCaMo* agents. JADE supplies standard agent services, i.e. mes-sage passing, distributed containers, naming and yellow pages services, agent mobility. When needed, an agent can enact a protocol role, which provides a set of operations by means of which agents participate in a mediated interaction session. JaCaMo [10] is a platform integrating Jason (as an agent programming language), CArtAgO and Moise (as a support to the realization of organizations). Normative/organizational specification is expressed as a Moise organization and translated into artifacts, that agents can decide to use.

We realize commitment-based interaction protocols by means of CArtAgO [21] artifacts. The core of 2COMM is in charge of management, maintenance and update of the social interaction state associated to each instance of a pro-tocol artifact. CArtAgO provides a way to define and organize *workspaces*, that are logical groups of artifacts, that can be joined by agents at runtime. The environment is itself programmable and encapsulates services and functionali-ties. An API allows programming *artifacts*, regardless of the agent programming language or the agent framework used. This is possible by means of the *agent body* metaphor: CArtAgO provides a native agent entity, which allows using the framework as a complete MAS platform as well as it allows mapping the agents of some platform onto the CArtAgO agents, which, in this way, becomes a kind of "proxy" in the artifacts workspace. The former agent is the mind, that uses the CArtAgO agent as a body, interacting with artifacts. An agent interacts with an artifact by means of public *operations*, which can be equipped with *guards*: conditions that must hold in order for operations to produce their effects.

2COMM is organized as follows. Protocol roles are provided by *communica-tion artifacts*, that are implemented by means of CArtAgO. Each communication artifact corresponds to a specific protocol enactment and maintains an own social state and an own communication state. Roles are linked to agents of the specific platform, via connector classes that implements the `IPlayer` interface. Figure 1 reports an excerpt of the 2COMM UML class diagram[1]. Let us get into the depths of the implementation:

---

[1] The source files of the system and examples are available at the URL http://di. unito.it/2COMM.

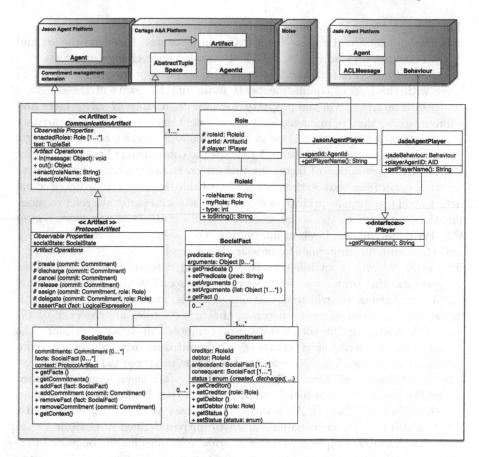

Fig. 1. Excerpt of the UML class diagram of 2COMM and connectors for JADE and JaCaMo.

- *CommunicationArtifact* (CA for short) provides the basic communication operations *in* and *out* for allowing mediated communication. CA extends an abstract version of the *TupleSpace* CArtAgO artifact: briefly, a blackboard that agents use as a tuple-based coordination means. In and out are, then, operations on the tuple space. CA also traces who is playing which role by using the property *enactedRoles*.
- Class *Role* extends the CArtAgO class *Agent*, and contains the basic manipulation logic of CArtAgO artifacts. Thus, any specific role, extending this super-type, will be able to perform operations on artifacts, whenever its player will decide to do so. Role provides static methods for creating artifacts and for *enacting/deacting* roles; the connector is in charge of linking agent and protocol through an instance of requested role.

- The class *CARole* is an inner class of CA and extends the Role class. It provides the *send* and *receive* primitives, implemented based on the *in* and *out* primitives provided by CA, by which agents can exchange messages.
- *ProtocolArtifact* (PA for short) extends CA and allows modeling the social layer with the help of commitments. It maintains the state of the on-going protocol interaction, via the property *socialState*, a store of social facts and commitments, that is managed only by its container artifact. This artifact implements the operations needed to manage commitments (create, discharge, cancel, release, assign, delegate). PA realizes the commitment life-cycle and for the assertion/retraction of facts. Operations on commitments are realized as *internal operations*, that is, they are not invokable directly: the protocol social actions will use them as primitives to modify the social state. We refer to modifications occurred to the social state as *social events*. Being an extension of CA, PA maintains two levels of interaction: the social one (by commitments), and the communication one (by message exchange).
- The class *PARole* is an inner class of PA and extends the CARole class. It provides the primitives for *querying the social state*, e.g. for asking the commitments in which a certain agent is involved, and the primitives that allow an agent to become, through its role, an *observer of the events* occurring in the social state. For example, an agent can query the social state to verify if it contains a commitment with a specific condition as consequent, via the method `existsCommitmentWithConsequent` (`InteractionStateElement el`). Alternatively, an agent can be notified about the occurrence of a social event, provided that it implements the inner interface *ProtocolObserver*. Afterwards, it can start observing the social state. PARole also inherits the communication primitives defined in CARole.
- The class *SocialFact* represents a fact of some relevance for the ongoing interaction, that holds in the current state of interaction. A social fact is asserted for tracking the execution of a protocol action. Actions can have additional effects on the social state; in this case, corresponding social facts are added to it.
- The class *IPlayer* is the interface between roles and players adopting them. Currently 2COMM provides implementations for Jade (*JadeBehaviourPlayer*) and Jason (*JasonAgentPlayer*).

In order to specify a *commitment-based interaction protocol*, it is necessary to extend PA by defining the proper social and communicative actions as operations on the artifact itself. Since we want agents to act on artifacts only through their respective roles, when defining a protocol it is also necessary to create the roles. We do so by creating as many extensions of PARole as protocol roles. These extensions are realized as inner classes of the protocol: each such class will specify, as methods, the powers of a role. Powers allow agents who play roles to actually execute artifact operations. The typical schema will be:

```
1  public class MyProtocolArtifact
2    extends ProtocolArtifact {
3    // ...
4    static {
5      addEnabledRole("Role1", Role1.class);
6      addEnabledRole("Role2", Role2.class);
7      // ...
8    }
9    // MY PROTOCOL ARTIFACT OPERATIONS
10   @OPERATION
11   @ROLE(name="roleName")
12   public void op1(...) {
13     // prepare a message, if needed; in that case,
14     send(message);
15     // modify the social state,
16     // e.g. create commitment, update commitment
17   }
18   // ...
19   // INNER CLASSES for ROLES
20   public class Role1 extends PARole {
21     public Role1(Behaviour player, AID agent) {
22       super("Role1", player, agent);
23     }
24     // define social actions for Role1
25     public void action1(...) {
26       doAction(this.getArtifactId(),
27         new Op("op1", ..., getRoleId()));
28     }
29     // ...
30   }
31   public class Role2 extends PARole {
32     // ...
33   }
34   // ...
35 }
```

Protocol designers program the interaction protocol once. The resulting artifact can, then, be used in a JADE or in a JaCaMo context.

3.1   Agent Programming with 2COMM

Agent programming with 2COMM amounts, at its core, to realizing a classical "sense-plan-act cycle", whose phases can be renamed "behold the social state", "activate behaviors according to social events", "schedule behavior execution". Beholding the social state, in 2COMM, does not require the agent to proactively sense it because agents can register to the protocol artifacts they use, and be notified by such artifacts of the occurrence of all events that are socially relevant. Agent programmers need to implement behaviors for those social events (e.g. commitment creation, commitment detachment) their agents are expected to handle. The occurrence of a social event for which an agent has a behavior to execute causes the activation of that behavior that will, then, be scheduled for execution by the agent platform. This mechanism represents an agent-oriented declination of callbacks. The agent paradigm forbids to use pure method invocation on the agent, that is autonomous by definition. Instead, the agent designer provides a collection of behaviours in charge of handling the different, possible evolutions of the social state, that are scheduled for execution when the corresponding conditions occur. For instance, when an agent has a behavior whose execution can create a commitment (the agent will be the debtor of that commitment), it needs to be able to tackle all events involved in the life cycle of that commitment (e.g. its detachment, satisfaction, violation, etc.). Instead, when, along the protocol execution, a conditional commitment may be created, whose creditor is the agent being implemented, then, a behavior that causes the detachment of the commitment should be included in the agent implementation.

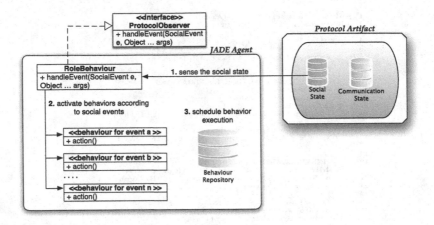

Fig. 2. 2COMM event handling schema for JADE agents.

Behaviors for tackling the satisfaction/violation should also be considered. In this way, an intuitive programming guideline, based on social events, is provided to agent developers.

Programming JADE Agents with 2COMM. A JADE agent which plays a role in some protocol artifact needs to implement the method `handleEvent` in its behaviors (Fig. 2). When events occur, the protocol artifact notifies them to the agents focusing on it. These react to such events by way of `handleEvent`. The agent programmer needs just to implement the logic for handling them, usually by adding proper behaviour(s) to the agent's behaviour repository. The occurrence of an event activates a corresponding behavior, if any is defined. When scheduled, the behaviour will be executed, and the event handled. The following is the pseudo-code of an example implementation:

```
1  public class MyBehaviour extends
2     SomeJadeBehaviour implements ProtocolObserver {
3     [ ... ]
4     public void action() {
5        ArtifactId art = Role.createArtifact
6           (myArtifactName, MyArtifact.class);
7        myRole = (SomeRole) (Role.enact
8           (MyArtifact.ROLE_NAME, art,
9              new JadeBehaviourPlayer(this, myAgent.getAID())));
10       myRole.startObserving(this);
11       // add the initial behaviour of the agent
12    }
13    public void handleEvent(SocialEvent e,
14                Object... args) {
15       SETemplate t = new SETemplate(myRole.getRoleId());
16       SETemplate t2 = new SETemplate(myRole.getRoleId());
17       t.iAmDebtor().commitIsDetached().consequentMatch(...);
18       t2.iAmCreditor().commitIsConditional().antecedentMatch(...);
19       if (t.match(e) {
20          myAgent.addBehaviour(...);   // behaviour to handle the case
21       } else if (t2.match(e)) {
22          myAgent.addBehaviour(...); // behaviour to handle another case
23       } else
24          // ...                      // behaviours for different cases
25       }
26    }
27 }
```

The basic schema, proposed for implementing a JADE behaviour, tracks how to handle social events that a protocol artifact notifies to an agent. Notification is performed through the *handleEvent* method, whose parameter contains the social effects of the event (e.g. if a commitment is added or satisfied, if a social fact is asserted, and such like). The implementation of *handleEvent* should contain conditions related to the occurred event. In JADE, event-related behaviours are added to the agent's behaviour library when a certain condition holds. If the social event to be notified is a commitment, it is possible to further check specific conditions of interest on it, including its state, the identity of its debtor and/or creditor, the antecedent or consequent condition (lines 19–22). The agent will, then, add appropriate behaviours to handle the detected situation. A template-based matching mechanism for social events is provided (class `SETemplate`, lines 15–18) used by programmer in order to specify matching conditions. Each template class method returns `this`, thus compacting the code for construction of complex conditions simply using the standard method dot notation.

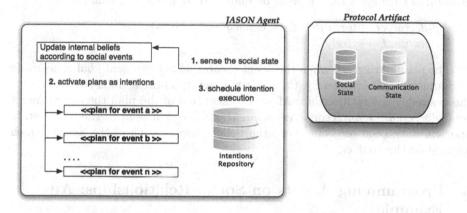

Fig. 3. 2COMM event handling schema for JaCaMo/Jason agents.

Programming JaCaMo Agents with 2COMM. JaCaMo agents are programmed as Jason agents. A Jason agent has the capability of performing reasoning cycles, that is, the agent architecture performs a cycle of sense-plan-act, that allows agents to evaluate which plans are triggered for execution each time an event occurs. In this framework, social events can be modeled as regular Jason events, fired by the protocol artifact, thus, it is not required to the Jason agents to perform any processing of them. This is a difference with JADE, which, instead, provides the abstraction of agent only as a set of behaviours with communication capabilities. The adoption of artifacts that notify the occurrence of social events to focusing agents allows plan specifications whose triggering events involve social events (e.g. commitment creation), as depicted in Fig. 3. Social events can also be used inside a plan context or body. As a difference with beliefs, commitment assertion/deletion can only occur through the artifact, as a consequence of a modification of the social state. For example, this is the case that deals with commitment addition:

$+cc(debtor, creditor, antecedent, consequent, status)$:
    $\langle context \rangle \leftarrow \langle body \rangle$.

The plan is triggered when a commitment that unifies with the plan head appears in the social state with the specified status. The syntax is the standard for Jason plans. *Debtor* and *Creditor* are to be substituted by the proper role names. The plan may be devised so as to achieve a change of the status of the commitment (e.g.: the debtor will satisfy the consequent, the creditor will satisfy the antecedent and so detach the commitment) or it may be devised to allow the agent to do something as a reaction (e.g. collecting information). A similar schema can be used in the case of commitment deletion and in the case of addition (deletion) of social facts. Commitments can also be used in contexts and in plans as test goals ($?cc(\dots)$) or achievement goals ($!cc(\dots)$). Addition or deletion of such goals can, as well, be managed by plans. For example:

$+!cc(debtor, creditor, antecedent, consequent, status)$:
    $\langle context \rangle \leftarrow \langle body \rangle$.

The plan is triggered when the agent creates an achievement goal concerning a commitment. Consequently, the agent will act upon the artifact so as to create the desired social relationship. After the execution of the plan the commitment $cc(debtor, creditor, antecedent, consequent, status)$ will hold in the social state and will have been projected onto the belief bases of each of the agents which focused on the artifact.

4   Programming Agents on Social Relationships: An Example

2COMM protocols constitute an outline for building agents that entails a uniform implementation of agent social abilities among different platforms: how to react to social events. We present a real implementation of the Contract Net Protocol (CNP) [17], and how JADE and Jason agents can be implemented. We adopt the following commitment-based CNP formulation:

cfp **causes** create($C(i, p, propose, accept \vee reject)$)
$accept$ **causes** *none*
$reject$ **causes** release($C(p, i, accept, done \vee failure)$)
$propose$ **causes** create($C(p, i, accept, done \vee failure)$)
$refuse$ **causes** release($C(i, p, propose, accept \vee reject)$)
$done$ **causes** *none*
$failure$ **causes** *none*

where i stands for the role *Initiator* and p for *Participant*. The execution of each such action has a social effect amounting to make the commitments which involve it progress, e.g. *propose* detaches *C(i, p, propose, accept ∨ reject)*. Additionally, they may cause further social effects which are those specified after "causes" (*none* means that the action execution has no further social effect but its occurrence). Initiator supplies its player the actions *cfp* (call for proposal), *accept*, and *reject*. The first allows the initiator to ask participants for proposals for solving a task of interest. If a proposal is chosen, action *accept* notifies the winner and all other proposals are rejected. The role participant supplies its player the actions *propose*, *refuse*, *done*, and *failure*. Action *propose* allows a participant to supply a solution for a task, action *refuse* allows declining the invitation to send a proposal. If a proposal is accepted, the winning participant is expected to execute the task and either provide the result by means of the action *done* or communicate its failure. Actions affect the social state, e.g., when an Initiator executes *cfp*, the commitment *C(i, p, propose, accept ∨ reject)* is added to the social state. This binds *i* to either accept or reject a proposal, if one is received.

```
1  public class Cnp extends ProtocolArtifact {
2    private int numberMaxProposals = 10;
3    private int actualProposals = 0;
4    // ... other protocol operations ...
5    @OPERATION
6    @ROLE(name="participant")
7    public void propose(String prop, int cost,
8                        String init) {
9      Proposal p = new Proposal(prop, cost);
10     RoleId participant = getRoleIdByPlayerName(getOpUserName());
11     RoleId initiator = getRoleIdByRoleCanonicalName(init);
12     p.setRoleId(participant);
13     RoleMessage propMessage = new RoleMessage(
14         participant, initiator, ACLMessage.PROPOSE, proposal);
15     send(propMessage);
16     defineObsProperty("proposal",
17         p.getProposalContent(), p.getCost(),
18         participant.getCanonicalName());
19     createCommitment(new Commitment(participant,
20         initiator, "accept", "done OR failure"));
21     assertFact(new Fact("propose", participant,
22                         prop));
23     actualProposals++;
24     if (actualProposals == numberMaxProposals) {
25       RoleId groupParticipant = new RoleId("participant");
26       createCommitment(new Commitment(initiator,
27                     groupParticipant,
28                     "true", "accept OR reject"));
29     }
30     // ... other protocol operations ...
31     // Role classes
32   public class Initiator extends PARole {
33     // ...
34     public void cfp(Task task) {
35       doAction(this.getArtifactId(), new Op("cfp", task, getRoleId()));
36     }
37     // ...
38   }
39   public class Participant extends PARole {
40     public void propose(Proposal proposal, RoleId proposalSender) {
41       // ...
42     }
43     // ...
44   }
45 }
```

The *propose* (line 7) is a social action. It is realized as a CArtAgO operation, in fact it is decorated by the CArtAgO Java annotation @OPERATION, line 5. It can be executed only by an agent playing the role *participant*. This is specified by the 2COMM Java annotation @ROLE(name="participant"), line 6. It asserts social fact (line 21), that traces the proposal made by the participant; then, it counts the received proposals and, when their number is sufficient, signals this fact to the initiator by the creation of a commitment (line 19) towards the group of participants. A message of performative PROPOSE (line 13) containing the participant's proposal is sent to the initiator.

The proposed CNP implementation remains the same independently from the fact that it is used by a JADE or a Jason agents. 2COMM current version uses role internal classes for JADE agents, while these are ignored if the enacting agents are written in Jason. Let us now compare agents implementations to highlight similarities and analogies. We will focus on the code for the Initiator role, starting from a JADE agent.

JADE Agents. Protocol designers can provide full support to JADE developers by implementing *social-event adapters* (see the listing below). So, agent developers will only need to implement the specific behaviours that tackle the events that are notified by the artifact. A clear advantage is an increased code reuse and modularization: the agent needs to be able to react to social events, adopting corresponding behaviours, and, therefore, the agent's autonomy is not jeopardized by extending the adapter. Here is a possible implementation for the Initiator adapter behaviour.

```
1  public abstract class InitiatorAdapterBehaviour
2          extends OneShotBehaviour
3          implements ProtocolObserver {
4   public String artifactName;
5   protected Initiator initiator;
6   public abstract Behaviour
7       commitToAcceptOrRejectIfPropose();
8   public abstract Behaviour
9       satisfyCommitToAcceptOrReject();
10  public abstract Behaviour
11      fulfilledCommitToDoneOrFailure();
12  public InitiatorBehaviour(String artifactName){
13      this.artifactName = artifactName;
14  }
15  public void action() {
16      ArtifactId art = Role.createArtifact(artifactName,
17          CNPArtifact.class);
18      initiator = (Initiator) (Role.enact(
19          CNPArtifact.INITIATOR_ROLE, art, this,
20          myAgent.getAID()));
21      initiator.startObserving(this);
22      myAgent.addBehaviour(
23          this.commitToAcceptOrRejectIfPropose());
24  }
25  public void handleEvent(SocialEvent e,
26      Object... args) {
27      SETemplate t = new SETemplate(initiator.getRoleId());
28      t.iAmDebtor().commitIsDetached();
29      t.matchCreditor(CNPArtifact.PARTICIPANT_ROLE);
30      t.matchConsequent("accept OR reject");
31      if (t.match(e)) {
32          myAgent.addBehaviour(satisfyCommitToAcceptOrReject());
33      } else {
34          t.matchConsequent("done OR failure");
35          if (t.match(e))
36              myAgent.addBehaviour(fulfilledCommitToDoneOrFailure());
37      }
38  }}
```

After line 21, all events, occurring in the social state, are notified to the role Initiator, which will handle them by executing *handleEvent* after a callback. The above abstract behaviour is extended by the concrete behaviour of the agent that plays the role Initiator. In particular, here we find the methods that create the actual behaviours for managing the social events.

```
1   public class InitiatorAgent extends Agent {
2     // ...
3     public class InitiatorBehaviourImpl
4         extends InitiatorBehaviour {
5       public final String ARTIFACT_NAME = "CNP-1";
6       public InitiatorBehaviourImpl() {
7         super(ARTIFACT_NAME);
8       }
9       public Behaviour commitToAcceptOrRejectIfPropose(){
10        return new CommitToAcceptOrRejectIfPropose(
11          initiator);
12      }
13      public Behaviour satisfyCommitToAcceptOrReject(){
14        return new SatisfyCommitToAcceptOrReject(initiator);
15      }
16      public Behaviour fulfilledCommitToDoneOrFailure(){
17        return new FulfilledCommitToDoneOrFailure(initiator);
18      }
19    }
20  }
```

The agent logic is structured as a number of behaviours that are in charge of handling the social events. When a social event is received, the adapter loads the corresponding behaviour, that is scheduled for the execution. This is similar to how a Jason agent is programmed, that is, a collection of plans that become active when a trigger is satisfied. For example, we report the behaviour *SatisfyCommitToAcceptOrReject*, which gathers proposals and selects the one to accept[2].

```
1   public class SatisfyCommitToAcceptOrReject
2       extends OneShotBehaviour {
3     Initiator initiator = null;
4     ArrayList<Proposal> proposals =
5       new ArrayList<Proposal>();
6     public SatisfyCommitToAcceptOrReject(
7         Initiator initiator) {
8       super();
9       this.initiator = initiator;
10    }
11    public void action() {
12      ArrayList<RoleMessage> propos =
13        initiator.receiveAll(ACLMessage.PROPOSE);
14      for (RoleMessage p : propos) {
15        proposals.add((Proposal) (p.getContents()));
16      }
17      initiator.accept(proposals.get(0));
18      for (int i = 1; i < proposals.size(); i++) {
19        initiator.reject(proposals.get(i));
20      }
21    }
22  }
```

JaCaMo Agents. The above implementation of *SatisfyCommitToAcceptOrReject* is analogous to how a Jason agent can be programmed to react to the same commitment:

```
1
2   +cc(My_Role_Id, "participant", "true",
3             "(accept OR reject)", "DETACHED")
4   : enactment_id(My_Role_Id) & not evaluated
5   <- +evaluated;
6      .wait(2000);
7      .findall(proposal(Content,Cost,Id),
8               proposal(Content,Cost,Id),Proposals);
9      .min(Proposals,proposal(Proposal,Cost,Winner_Role_Id));
10     accept(Winner_Role_Id).
11     ... action 'reject' for all other proposals ...
```

---

[2] The criterion here is trivial, the first proposal is the one which is accepted.

We now report and comment excerpts of Jason agent code for the *Initiator*.

```
1  /* Initial goals */
2  !startCNP.
3  /* Plans */
4  +!startCNP : true
5      <- makeArtifact("cnp","cnp.Cnp",[],C);
6         focus(C);
7         enact("initiator").
8  +enacted(Id,"initiator",Role_Id)
9      <- +enactment_id(Role_Id);
10        !cc(Role_Id, "participant", "propose",
11            "(accept OR reject)","CONDITIONAL").
12 +!cc(My_Role_Id, "participant", "propose",
13           "(accept OR reject)","CONDITIONAL")
14     <- .print("sending cfp");
15        .wait(2000);
16        cfp("task-one").
17 +cc(My_Role_Id, "participant", "true",
18           "(accept OR reject)", "DETACHED")
19     :  enactment_id(My_Role_Id) & not evaluated
20     <- +evaluated;
21        .wait(2000);
22        .findall(proposal(Content,Cost,Id),
23              proposal(Content,Cost,Id),Proposals);
24        .min(Proposals,proposal(Proposal,Cost,Winner_Role_Id));
25        accept(Winner_Role_Id).
26        ... action 'reject' for all other proposals ...
27 +cc(Participant_Role_Id, My_Role_Id, "true",
28           "(done OR failure)", "DISCHARGED")
29     :  done(Result)
30     <- .print("Task resolved: ",Result).
31 +cc(Participant_Role_Id, My_Role_Id, "true",
32           "(done OR failure)", "DISCHARGED")
33     :  failure(Participant_Role_Id)
34     <- .print("Task failed by ",Participant_role_id).
```

!startCNP, line 2, is an initial goal, that is provided for beginning the interaction. In this implementation, the agent which plays the initiator role is in charge of creating the artifact (*makeArtifact("cnp", "cnp.Cnp", [], C)*) that will be used for the interaction. The agent will, then, enact the role "initiator" (*enact("initiator")*); the artifact will notify the success of the operation by asserting an *enacted* belief. Since the program contains the plan triggered by the *enacted* belief, the initiator agent can, then, execute *cfp*. When enough participants will have committed to perform the task, in case their proposal is accepted (*cc(My_Role_Id, "participant", "true", "(accept OR reject)", "DETACHED")*), the initiator agent evaluates the proposals and decides which to *accept* (we omit the reject case for sake of brevity).

Table 1. Comparison among JADE, JaCaMo and 2COMM improvements.

	JADE	+ 2COMM	JaCaMo	+ 2COMM
Programmable communication channels	X	✓	✓	✓
Notification of social relationships of interaction	X	✓	X	✓
Interaction/agent logic decoupling	X	✓	X	✓
Expected behaviours reasoning	X	✓	✓	✓
Library of reusable patterns of interaction	✓	✓	X	✓
Runtime interaction monitoring	X	✓	✓	✓
Social-based Agent Programming Pattern	X	✓	X	✓
Norms and Obligations modeling	X	X	✓	✓

Remarks. Summarizing, a JADE agent leveraging 2COMM artifacts consists of a set of behaviours aimed at accomplishing given social relationships: such behaviours depend neither on when nor on how the social relationships of interest are created inside the social state. These aspects are, in fact, encoded in the protocol artifact that creates them based on the actions the agents perform. As a consequence, modifying how or when a social relationship is created does not have any impact on the agent implementation. Analogously for Jason agents, plans are not affected by modifications made on protocol: it is possible to adapt the interaction logic to different contexts without any impact on agents. Each plan is defined as reaction to a social event, whose evolution is stated by the artifact. Table 1 synthesizes a comparison among JADE and JaCaMo, highlighting aspects that are improved or added by 2COMM.

5 Conclusions and Discussion

In this work, we have proposed 2COMM, an infrastructure for allowing agents to interact by following an accepted set of regulations, with a self-governance approach. Self-governance mechanisms rely on the reification of commitments and of commitment-based protocols. These are, at all respects, resources that are made available to stakeholders and that are realized by means of artifacts. The proposal is characterized, on the one hand, by the flexibility and the openness that are typical of MAS, and, on the other, by the modularity and the compositionality that are typical requirements of the methodologies for design and development. One of the strong points of the proposal is the decoupling between the design of the agents and the design of the interaction, that builds on the decoupling between computation and coordination done by coordination models, like tuple spaces. This is a difference with respect to JADE or JaCaMo where no decoupling occurs: a pattern of interaction is projected into a set of JADE behaviours or Jason plans, one for each role. Binding the interaction to ad-hoc behaviours/plans does not allow having a global view of the protocol and complicates its maintenance.

2COMM supports programming JADE and Jason agents, by following a uniform approach. Moreover, agents programming can leverage the methodology explained in [3]. It also simplifies the interaction of different agent platforms thanks to its connectors, which make protocol artifacts accessible and, thus, allow mediated communication between agents that belong to different platforms. This feature fulfills the purpose of supporting the development of heterogeneous and open agent systems. So, for instance, 2COMM enables the interaction of JADE agents with JaCaMo agents in a transparent and seamless way: it is not necessary to adapt an agent implementation to the platforms on which the other agents of the system run. For what concerns implementation, connectors bridge between CArtAgO and the used agent platforms. Any agent can take part to interaction sessions with others simply by using a protocol artifact. Note that Jason (which is part of JaCaMo) allows changing the communication infrastructure, switching to that of JADE, however, this choice is to be done a priori and has an impact on the design. 2COMM does not impose any choice a priori, guaranteeing the interaction between any pair of agent platforms for which a 2COMM connector exists.

2COMM complements the obligation-based specification of organizations, specifically suiting those situations where interaction is not subject to an organizational guideline, like in the case when interaction is among agents and each agent decides what is best for itself, or when guidelines amount to declarative, underspecified constraints that still leave agents the freedom to take strategic decisions on their behavior. In this case, interaction strongly relies on the two basic notions of goal and of engagement. For a thorough discussion of the differences between our proposal and organizational or normative approaches, please check [4].

Decoupling is an effect of explicitly representing social relationships as resources: agent behaviour is, thus, defined based on the existing social relationships and not on the process by which they are created. For instance, in CNP the initiator becomes active when the commitments that involve it as a debtor, and which bind it to accept or reject the proposals, are detached. It is not necessary to specify nor to manage, inside the agent, such things as deadlines or counting the received proposals: the artifact is in charge of these aspects. Testing 2COMM with Jason and JADE proved that programming agents starting from their desired interaction can be a valuable starting point, that can be extended towards a methodology useful for open and heterogeneous scenarios. We intend to explore this direction by adding connectors for different agent platforms.

Recently, we developed on top of 2COMM a commitment-based typing system [2] for JADE agents. Such typing includes a notion of compatibility, based on subtyping, which allows for the safe substitution of agents to roles along an interaction that is ruled by a commitment-based protocol. Type checking can be done dynamically when an agent enacts a role.

References

1. Baldoni, M., Baroglio, C., Capuzzimati, F.: A commitment-based infrastructure for programming socio-technical systems. ACM Trans. Internet Technol. (TOIT) **14**(4), 23 (2014)
2. Baldoni, M., Baroglio, C., Capuzzimati, F.: Typing multi-agent systems via commitments. In: Dalpiaz, F., Dix, J., van Riemsdijk, M.B. (eds.) EMAS 2014. LNCS, vol. 8758, pp. 388–405. Springer, Heidelberg (2014)
3. Baldoni, M., Baroglio, C., Capuzzimati, F., Micalizio, R.: Empowering agent coordination with social engagement. In: Gavanelli, M., Lamma, E., Riguzzi, F. (eds.) AI*IA 2015: Advances in Artificial Intelligence, XIV International Conference of the Italian Association for Artificial Intelligence. LNAI, Ferrara, Italy, September 2015. Springer (to appear)
4. Baldoni, M., Baroglio, C., Capuzzimati, F., Micalizio, R.: Leveraging commitments and goals in agent interaction. In: Ancona, D., Maratea, M., Mascardi, V. (eds.) Proceedings of XXX Italian Conference on Computational Logic, CILC 2015, Genova, Italy, July 2015

5. Baldoni, M., Baroglio, C., Mascardi, V., Omicini, A., Torroni, P.: Agents, multi-agent systems and declarative programming: what, when, where, why, who, how? In: Dovier, A., Pontelli, E. (eds.) GULP. LNCS, vol. 6125, pp. 204–230. Springer, Heidelberg (2010)
6. Baldoni, M., Boella, G., van der Torre, L.W.N.: Modelling the interaction between objects: roles as affordances. In: Lang, J., Lin, F., Wang, J. (eds.) KSEM 2006. LNCS (LNAI), vol. 4092, pp. 42–54. Springer, Heidelberg (2006)
7. Baldoni, M., Boella, G., van der Torre, L.: Interaction between objects in power-java. J. Object Technol. **6**(2), 5–30 (2007)
8. Bellifemine, F., Bergenti, F., Caire, G., Poggi, A.: JADE - a Java agent development framework. In: Bordini, R.H., Dastani, M., Dix, J., El Fallah Seghrouchni, A. (eds.) Multi-agent Programming: Languages, Platforms and Applications. Multiagent Systems, Artificial Societies, and Simulated Organizations, vol. 15, pp. 125–147. Springer, New York (2005)
9. Boella, G., van der Torre, W.N.: The ontological properties of social roles in multi-agent systems: definitional dependence, powers and roles playing roles. Artif. Intell. Law **15**(3), 201–221 (2007)
10. Boissier, O., Bordini, R.H., Hübner, J.F., Ricci, A., Santi, A.: Multi-agent oriented programming with JaCaMo. Sci. Comput. Program. **78**(6), 747–761 (2013)
11. Bordini, R.H., Braubach, L., Dastani, M., El Fallah-Seghrouchni, A., Gómez-Sanz, J.J., Leite, J., O'Hare, G.M.P., Pokahr, A., Ricci, A.: A survey of programming languages and platforms for multi-agent systems. Informatica (Slovenia) **30**(1), 33–44 (2006)
12. Brazier, F.M.T., Dunin-Keplicz, B.M., Jennings, N.R., Treur, J.: Desire: modelling multi-agent systems in a compositional formal framework. Int. J. Coop. Inf. Syst. **06**(01), 67–94 (1997)
13. Chopra, A.K.: Commitment alignment: semantics, patterns, and decision procedures for distributed computing. Ph.D. thesis, North Carolina State University, Raleigh, NC (2009)
14. Chopra, A.K., Singh, M.P.: An architecture for multiagent systems: an approach based on commitments. In: Proceedings of ProMAS (2009)
15. Conte, R., Castelfranchi, C., Dignum, F.P.M.: Autonomous norm acceptance. In: Papadimitriou, C., Singh, M.P., Müller, J.P. (eds.) ATAL 1998. LNCS (LNAI), vol. 1555, pp. 99–112. Springer, Heidelberg (1999)
16. Fisher, M., Bordini, R.H., Hirsch, B., Torroni, P.: Computational logics and agents: a road map of current technologies and future trends. Comput. Intell. **23**(1), 61–91 (2007)
17. Foundation for Intelligent Physical Agents. FIPA Specifications (2002). http://www.fipa.org
18. Mascardi, V., Martelli, M., Sterling, L.: Logic-based specification languages for intelligent software agents. TPLP **4**(4), 429–494 (2004)
19. Omicini, A., Ricci, A., Viroli, M.: Artifacts in the A&A meta-model for multi-agent systems. Auton. Agents Multi-agent Syst. **17**(3), 432–456 (2008)
20. Omicini, A., Zambonelli, F.: TuCSoN: a coordination model for mobile information agents. In: 1st International Workshop on Innovative Internet Information Systems (IIIS 1998), pp. 177–187. IDI - NTNU, Trondheim, Norway, 8–9 June 1998
21. Ricci, A., Piunti, M., Viroli, M.: Environment programming in multi-agent systems: an artifact-based perspective. Auton. Agents Multi-agent Syst. **23**(2), 158–192 (2011)
22. Singh, M.P.: An ontology for commitments in multiagent systems. Artif. Intell. Law **7**(1), 97–113 (1999)

23. Singh, M.P.: A social semantics for agent communication languages. In: Dignum, F.P.M., Greaves, M. (eds.) Issues in Agent Communication. LNCS, vol. 1916, pp. 31–45. Springer, Heidelberg (2000)
24. Weyns, D., Omicini, A., Odell, J.: Environment as a first class abstraction in multiagent systems. Auton. Agents Multi-agent Syst. **14**(1), 5–30 (2007)
25. Yolum, I., Singh, M.P.: Commitment machines. In: Meyer, J.-J.C., Tambe, M. (eds.) ATAL 2001. LNCS (LNAI), vol. 2333, pp. 235–247. Springer, Heidelberg (2002)

The Emergence of Norms via Contextual Agreements in Open Societies

George A. Vouros[✉]

Department of Digital Systems, University of Piraeus, Piraeus, Greece
georgev@unipi.gr

Abstract. This paper explores the emergence of norms in agents' societies when agents play multiple - even incompatible - roles in their social contexts simultaneously, and have limited interaction ranges. Specifically, this article proposes two reinforcement learning methods for agents to compute agreements on strategies for using common resources to perform joint tasks. The computation of norms by considering agents' playing multiple roles in their social contexts has not been studied before. To make the problem even more realistic for open societies, we do not assume that agents share knowledge on their common resources. So, they have to compute semantic agreements towards performing their joint actions.

1 Introduction

It is well known that effective norms, policies or conventions can significantly enhance the performance of agents acting in groups and societies, since they do enable a kind of social control to the behavior of agents, without compromising their autonomy [1]. The emergence or learning of norms in agents' societies is a major challenge, given that societies are open, agents may not be qualified to collaborate effectively under previously unseen conditions, or they may need to compute effective rules of behavior very efficiently, w.r.t. their preferences and constraints.

This article considers norms as social conventions (i.e. as set of agreed, stipulated, or generally accepted standards or criteria) rather than as deontic aspects (e.g. obligations, prohibitions or permissions). The main question that this article aims to answer is "how effectively do norms emerge in an open society via establishing agreements in social contexts through local interactions and with limited information about others knowledge, preferences and choices?".

Going beyond previous studies (e.g. [2–4]), this work aims at showing how agents mutually learn in a distributed and efficient way strategies that maximize their payoff w.r.t. some preferences and constraints (a) while playing multiple roles and maybe roles with incompatible preferences, (b) via local interactions in the social context, comprising multiple roles played by their acquaintances, in conjunction (c) to relaxing the assumption that agents share a common representation of the world. Towards this target, this article presents a formalization of the learning process and proposes two social reinforcement learning models where agents via communication with acquaintances in their social contexts

© Springer International Publishing Switzerland 2015
F. Koch et al. (Eds.): CARE-MFSC 2015, CCIS 541, pp. 185–201, 2015.
DOI: 10.1007/978-3-319-24804-2_12

form specific expectations for the behavior of others. The notion of agents' social context allows distinguishing between agreements of agents in their "local neighborhood" and society-wide norms. The above points clarify the contributions of this research work, compared to the state of the art approaches for learning norms.

Specifically, this article proposes methods for agents to reach social conventions, by considering the following novel features:

– Agents play multiple roles and interact with others that play multiple roles simultaneously in their social contexts.
– Agents reconcile conflicting options while considering incompatibilities among roles.
– Agents compute semantic correspondences between their representations of the world: This article deals with strategies on using resources towards performing joint tasks. Although the article considers a specific type of resource (time), this is not restrictive to the applicability of the methods proposed to other type of resources are strategies.

Finally,

– Agents learn society norms (conventions) via computing agreements in their social contexts.

We need to emphasize that agents' semantic agreements (i.e. agreements on the meaning of terms they use for the representation of resources) are put in the context of their joint tasks: Tasks that need the coordinated action of at least two agents. By doing so, agents are restricted to semantic agreements that do "work in reality" effectively. Consider for instance the case where agents, due to their limited knowledge of others' representations reach agreements about the meaning of symbols, which if put to a working context will lead either to ineffectiveness, or to the inability to act.

Towards answering the main research question stated above, this article proposes two social learning reinforcement learning methods. Both methods exploit (a) agents' preferences on the use of resources, (b) the feedback that agents receive for their strategies to use resources while interacting with others in their social context, (c) their reward for performing role-specific actions for any role they play w.r.t. their strategies for using resources. Although both methods are social (i.e. they are based on agents' local interactions), in one of them agents do not consider their joint decisions and do not share rewards, while in the other they learn collaboratively by acting and receiving rewards/sanctions for their joint strategies.

This article is structured as follows: Sect. 2 presents a motivating scenario. Section 3 formulates the problem and Sect. 4 presents the proposed methods. Section 5 presents experimental results for agents societies of different size and structure. Section 6 presents related works and finally, Sect. 7 concludes the article by summarizing the contributions made and presents interesting lines of future research.

2   Motivating Scenario

Consider the following scenario: AgentX among other commitments in its working context is being involved in a recently-appointed team towards the design of a new product. The team has a coordinator agent who has already many commitments with other teams. Besides that, AgentX is committed to perform other tasks as member of other groups independently from his working context. Role-specific tasks that AgentX has to perform require resources whose use has to be coordinated with others. Time is of primary importance here. Thus, to arrange his schedule, AgentX considers different time periods for scheduling role-specific tasks according to roles' preferences (e.g. due to conventional arrangements). He tries to keep concerns separate, while complying with his commitments and obligations and coordinating with others effectively.

The social context of AgentX is the set of roles to which the agent interacts, including also own roles. The agents with whom he interacts may also play multiple roles, and they constitute AgentX's neighborhood.

Consider for instance two daily periods that AgentX names P_1 and P_2. Given a role R that AgentX plays, these periods may be ordered according to a measure of preference: Let that be $\gamma(R, \cdot)$. Let for instance P_1 be more preferred than P_2 from the point of view of role R. I.e. $\gamma(R, P_2) \leq \gamma(R, P_1)$. While AgentX's collaborators may play other roles (different from R) they may not share AgentX' representation of periods. Thus, AgentX has to agree with them on the meaning of periods, to start coordinating with them. Consider for instance the periods X_1 and X_2 specified by the busy team coordinator. Agents, to begin coordinating towards performing joint tasks (e.g. meeting), need to reach agreement to the correspondences between the periods considered by the different roles[1]. In this particular case there are clearly two possible options for reaching an agreement on correspondences between periods. Nevertheless, there is at most one option which is meaningful (i.e. it corresponds to the semantics of periods' representations), but we do not assume this to be known to the agents. Let, for instance, the meaningful correspondences be: (a) P_1 is the same as X_2 and (b) P_2 is the same as X_1. Notice that agents may reach an agreement to non-meaningful correspondences: In this case they will not be able to act jointly, receiving a very low payoff for their joint task.

Notice that to reach an agreement to the correspondences between periods' representations agents do not have to consider their preferences on periods for scheduling tasks. Nevertheless, considering preferences, AgentX's neighbors may have the incentive of choosing AgentX's non-preferable period for scheduling their tasks. Such decisions can lead to undesirable situations and to ineffectiveness in performing joint tasks. Consider for instance the busy team coordinator. He/she may prefer X_1 to schedule meetings with team members, while AgentX-being a member of the team, prefers to schedule joint tasks in period P_1.

---

[1] Of course agents may specify periods using different time granularities, different forms of representing time, etc. In this paper we assume that there is a specific granularity for specifying periods and thus agents just have to align their specifications: Otherwise, further agreements are necessary.

Given, for instance, that P_1 is the same as X_2 and P_2 is the same as X_1, then, the possible choice of the team member to schedule a joint task is P_1 (and X_2), while for the team coordinator is X_1 (and P_2): These possible choices do not satisfy the preferences of both agents.

In addition to these, some of the roles may have incompatible requirements and preferences to the use of resources. We define two roles to be *incompatible w.r.t. a resource* (or simple *incompatible*, in case we consider time as the only type of resources) if joint tasks for these roles cannot share the resource when performed by a single agent (e.g. considering time periods, an agent must schedule tasks for two incompatible roles in non-overlapping time periods).

Thus, summarizing the above, AgentX has to reach agreements with his neighbors to schedule their joint tasks, so as to satisfy as much as possible his preferences on scheduling tasks, and the constraints related to the incompatibility of roles: This is rather complicated given that AgentX plays multiple roles and interacts with multiple others, while this is true for his acquaintances.

For a convention to evolve in the society, all agents in the population playing the same roles have to agree on their strategies for using resources: E.g. pairs of agents playing the roles of team members and team coordinators have to learn one of the following policy pairs to schedule joint tasks: (a) (P_1, X_2), according to the preference of the team member, or (b) (P_2, X_1), being in accordance to the preferences of the coordinator.

This scenario emphasizes on the following aspects of the problem:

– Related to resources:
  • Agents need to coordinate their use of resources to perform joint tasks (in our scenario we consider time as the unique resource).
  • Agents do not share a common representation of the resources, so they have to agree on the semantics of their representations.
  • Agents' preferences on the use of the resource vary for each of the roles they are playing.
– Related to agents' roles:
  • Each agent may play and interact with multiple (even incompatible) roles.
  • Each agent has a social context, defined by its own roles and the roles that it interacts with.
– Related to agreements and norms:
  • Semantic agreements are put in the context of agents actions: In our example it is clear that even if agents agree on correspondences between periods, this may not lead them to schedule their tasks as effectively as they may wish.
  • Agents in their social context have to reach agreements on the use of resources for performing their joint tasks.
  • Norms are agreements that are widely accepted by all agents in the society.

As far as we know, there is not any research work concerning the emergence of conventions in agents' societies that consider these aspects in combination. As already said, the major question that this paper aims to answer is "how effectively do norms emerge in a society via establishing agreements in social contexts through local interactions and with limited information about others' representations, preferences and choices?". The effectiveness of a model is measured by means of the percentage of role playing agents reaching agreement on specific conventions, as well as by measuring the computational iterations (epochs) necessary for a society to converge to conventions.

Towards answering this question, agents need to (a) compute semantic agreements for the terms they use to represent resources, (b) use semantic agreements to compute agreements on the use of resources for performing their joint tasks in their social contexts w.r.t. their preferences on using resources and roles' incompatibilities.

3 Problem Specification

A society of agents $S = (\mathcal{R}, A, E)$ is modeled as a graph with one vertex per agent in A and any edge in E connecting pairs of agents. A connected pair of agents must be coordinated to the use of resources for the performance of role-specific tasks (e.g. to the scheduling of their tasks) and can communicate directly to each other. Each agent i in the society is attributed with different roles $\mathcal{R} = \{R_1, R_2...\}$. The naming of roles is a social convention and thus, all agents in the society use the same set of roles. $N(i)$ denotes the neighborhood of agent i, i.e. the set of agents connected to agent i, including also itself. Subsequently, the fact that agent i plays the role $R_j \in \mathcal{R}$, is denoted by $i{:}j$.

Each role R_i considers a set of time periods $\mathcal{P}_{R_i} = \{P_1, P_2...\}$ that are ordered according to R_i's preferences for scheduling role-specific tasks. Role-specific periods in $\mathcal{P}_{R_i}$ are order by the preference of R_i, according to the function $\gamma(R_i, \cdot) : \mathcal{P}_{R_i} \to \mathbb{R}$. Although we may consider any relation between periods (e.g. they may be disjoint, overlapping etc.), in this article we consider only equal (=) and mutually disjoint (<>, non-overlapping) time periods. Each role has its own preferences to scheduling tasks in periods, while the naming of periods as well as the pairs of incompatible roles is common knowledge to all agents that play the same role.

Given a pair of roles (R_i, R_j), these may be *incompatible* w.r.t. a resource. Considering time, agents interacting with incompatible roles cannot schedule any pair of joint tasks, with each these roles, during the same time period. Any pair of agents, or a single agent, may play incompatible roles.

Agents playing different roles do not possess any common knowledge, neither exchange any information concerning the role-specific periods, their preferences on scheduling tasks, or their payoffs for scheduling tasks in any period. Thus, agents playing different roles may use different names for the same period, or the same name for denoting different periods. No agent possesses global knowledge on the semantics of role-specific representation of periods, and thus on correspondences between periods names: We consider that this holds for any single agent that plays multiple roles, as well.

At this point it must be emphasized that while this article considers time periods, the formulation and the proposed methods can be applied to other types of resources that can be treated similarly to time and are necessary to the execution of role-specific tasks.

A social context for an agent i denoted by $SocialContext(i)$, is the set of roles played by the agents in its neighborhood. More formally:

$$SocialContext(i) = \{R_k | \exists j \in N(i) \text{ and } j : k\}.$$

It must be noticed that the social context of an agent i includes own roles, denoted by $Roles(i)$.

Agents in the society must decide on the scheduling of their (more interestingly, joint) tasks so as to increase their effectiveness. More specifically, considering two acquaintances i:k and j:m, where $j \in N(i)$, and a joint task for their roles R_k and R_m, agents must schedule that task in an agreed period P, so as to increase their expected payoff with respect to their role-specific preferences on schedules. Considering that agents and their neighbors play multiple - maybe incompatible - roles, they have also to take into account role-specific (incompatible) requirements on scheduling tasks. Incompatibilities are formally specified in Sect. 3.

To agree on a specific period P for scheduling their joint task, agents i:k and j:m have to first agree on correspondences between their representations of periods: Towards this we consider that agents can subjectively hold correspondences between own representations of periods and representations of others: These may be computed by each agent using own methods, and information about others' roles. A subjective correspondence for the agent i:k and its acquaintance j:m is a tuple $\langle P, S \rangle$, s.t. $P \in \mathcal{P}_{R_k}$ and $S \in \mathcal{P}_{R_m}{}^2$. Such a correspondence represents that the agent i considers P and S to represent the same time interval. Nevertheless, given that acquaintances may nor agree on their subjective correspondences, they have to reach an agreed set of correspondences.

For norms to emerge in the society, any pair of agents (anywhere in the society) playing roles R_k, R_m must reach the same decisions for scheduling joint tasks for these roles.

Towards this goal, this article proposes two distributed social learning methods for agents to compute society-wide agreements via local interactions with their neighbors.

4   Social Reinforcement Learning Methods for Computing Agreements

To describe the proposed methods for the computation of norms, we distinguish between two, actually highly intertwined, computation phases: (a) The computation of agent-specific, subjective correspondences on periods, and strategies for scheduling tasks w.r.t. own preferences and constraints concerning incompatibility of roles; and (b) the computation of contextual agreements concerning agents' strategies to schedule joint tasks.

---

^{2} It must be pointed out that since the neighborhood of any agent includes itself, and its social context includes its own roles, it may also hold that $i = j$.

Computation of Local Correspondences and Strategies: Given an agent i playing a role R_k, and a role $R_m \in SocialContext(i)$ played by a an agent j in the neighborhood of i, agents need to compute subjective correspondences between periods in $\mathcal{P}_{R_k}$ and $\mathcal{P}_{R_m}$.

Although agents may use own methods to compute these correspondences, these computations have to preserve the semantics of periods' specifications: This is done via validity constraints that coherent correspondences between periods must satisfy. These constraints depend on the possible relations between periods. Therefore, considering only equal and disjoint time periods, and given two distinct roles R_k and R_m, the validity constraints that correspondences computed by $i{:}k$ must satisfy are as follows:

- if $\langle P, X \rangle$ and $\langle P', X' \rangle$ are correspondences with $X, X' \in \mathcal{P}_{R_m}$, $P, P' \in \mathcal{P}_{R_k}$ and $P <> P'$, then it must hold that $X <> X'$.
- if $\langle P, X \rangle$ and $\langle P, X' \rangle$ are correspondences with $X, X' \in \mathcal{P}_{R_m}$ and $P, P' \in \mathcal{P}_{R_k}$, then $X = X'$.

Given these validity constraints, each agent can compute its own role-specific, subjective, coherent correspondences between time periods. Given these correspondences, any agent $i{:}k$ has to make a specific decision for the period to schedule joint tasks with any other agent playing the role R_m in its social context. Let that decision be denoted by $decision(i{:}k, \cdot{:}m)$[3]. Later on we specify how agents reach these decisions and how they reach agreements on their subjective correspondences.

Given that each agent may interact with multiple roles in its social context, considering any pair of incompatible roles R_k, R_m, the following incompatibility constraint holds:

- Given an agent i playing any role R_x, and given two incompatible roles $R_k, R_m \in SocialContext(i)$, then
  $decision(i : x, \cdot{:}m) <> decision(i : x, \cdot{:}k)$.

Given the above validity and incompatibility constraints, the utility of an agent $i : k$ for choosing a period $P \in \mathcal{P}_{R_k}$ to schedule joint tasks with $j : m$, given the subjective correspondence $< P, X >$ between periods, is $U(i{:}k, P) = \gamma(R_k, P) + f(i{:}k, P)$, where $\gamma(R_k, P)$ is the preference of role R_k to P, and $f(i{:}k, P) = G(i{:}k) + C(i{:}k)$, where $G(i{:}k) = Payoff * SatisfiedConstraints(i{:}k)$ and $C(i{:}k) = Penalty * ViolatedConstraints(i{:}k)$.

$Payoff$ is a positive number representing the payoff of any satisfied constraint in the social context of agent $i{:}k$ and $Penalty$ is a negative number that represents the cost of violating a validity or incompatibility constraint. $SatisfiedConstraints(i{:}k)$ (resp. $ViolatedConstraints(i{:}k)$) is the number of satisfied (resp. violated) constraints for the agent i.

---

[3] the notation $(\cdot{:}m)$ means "any agent playing the role R_m".

Computing Contextual Agreements: Given agents' subjective correspondences and own decisions for any role they play, these correspondences and decisions may not agree with the choices of their neighbors. Towards reaching agreements, also with respect to constraints and role-specific preferences, agents consider the feedback received from their neighbors.

According to this communication-based learning approach, given an agent i and two roles $R_k \in Roles(i)$ and $R_m \in SocialContext(i)$, to get feedback on decisions, the agent $i{:}k$ propagates its decision for scheduling joint tasks with agents $\cdot{:}m$ in its neighborhood in period P, together with its subjective correspondence $\langle P, X \rangle$, where $X \in \mathcal{P}_{R_m}$ to all R_m-playing agents in $N(i)$. It must be noticed that the propagated decision concerns a specific pair of role playing agents and both, a period and a subjective correspondence for this period. Such a decision is of the form $(i{:}k, x{:}m, \langle P, X \rangle)$, where $x \in N(i)$ and $decision(i{:}k, \cdot{:}m) = P$.

Agents propagate their decisions to their neighbors in the network iteratively and in a cooperative manner, aiming to exploit the transitive closure of correspondences in cyclic paths. This is similar to the technique reported in [5]. Agents propagate what we call *c-histories*, which are ordered lists of decisions made by agents along the paths in the network. Each propagated decision heads such a history. For instance the c-history propagated by i to any R_m-playing agent x, as far as the role R_k is concerned, is $[(i{:}k, x : m, \langle P, X \rangle)|L]$, where L is either an empty c-history or the c-history that has been propagated to i, concerning its role R_k. By propagating c-histories, agents can detect cycles and take advantage of the transitivity of correspondences, detecting positive/negative feedback to their decisions.

Specifically, an agent i detects a cycle by inspecting in a received c-history the most recent item $(i{:}k, x : m, \langle P, X \rangle)$ originated by itself: Given a cycle $(1 \rightarrow 2 \rightarrow ...(n-1) \rightarrow 1)$, then for each decision $(1{:}k, 2{:}m, \langle P, X \rangle)$ for the roles R_k and R_m that agents $1, 2$ play, respectively, heading a c-history from 1 to 2, the originator must get a decision $(n\text{-}1{:}m, 1{:}k, \langle P, X \rangle)$ from the last agent $(n - 1)$ in the cycle, if it plays the role R_m. Thus, the agent 1 must receive a decision from $(n - 1)$ concerning P, rather than to any other period, and the correspondence $\langle P, X \rangle$. In such a case the agent 1 counts a positive feedback. In case there is a cycle but the forwarded decision does not concern P, then there are one or more correspondences or decisions through the path that result to disagreements. In this case, the agent 1 counts a negative feedback for its decision. It must be noticed that disagreements may still exist when the agent 1 gets the expected choice but several decisions along the path compensate "errors". These cases are detected by the other agents, as the c-history propagates in the network. To make the computations more efficient and in order to synchronize agents' decision making we consider that c-histories can be propagated up to 3 hops with repetitions: This means that given two neighbors i and j, any c-history starting from i (1st hop) shall be returned to this agent with the decision of j (2nd hop), and will return later to j with the new decision of i (3rd hop).

In the last hop the agent i will choose a strategy by considering also the feedback received from j, in conjunction with feedback from any other neighbor.

But how actually do agents compute decisions in their social context w.r.t. their preferences and constraints? Notice that decisions concern specific periods w.r.t. subjective correspondences. From now on, when we say *decisions* we mean exactly this combination: Thus when agents revise their decisions they may revise their subjective correspondences, or their strategies for scheduling tasks, or both.

Reinforcement Learning and the Emergence of Norms: Given that agents do not have prior knowledge about the effects of decisions made, this information has to be learned based on the rewards received (including feedback from others). Using the model of collaborative multiagent MDP framework [6,7] we assume:

- The society of agents $S = (\mathcal{R}, A, E)$.
- A time step $t = 0, 1, 2, 3, ...$
- A set of discrete state variables per agent-role $i{:}k$ at time t, denoted by $s^t_{(i:k),(\cdot:m)}$, where $i \in A$ and $R_m \in SocialContext(i)$. The state variable ranges to the set of possible correspondences between periods in P_{R_k} and periods in P_{R_m}. The local state s^t_i of agent i at time t is the tuple of the state variables for all roles played by i in combination with any role in its social context. A global state s^t at time t is the tuple of all agents' local states. The set *State* is the set of global states.
- A strategy for every agent-role $i{:}k$ and role $R_m \in SocialContext(i)$ at time t, denoted by $c^t_{(i:k),(\cdot:m)} = decision(i{:}k, \cdot{:}m)$. The local strategy for every agent i, denoted by c^t_i is a tuple of strategies, each for any role that i plays in combination with any other role in its social context. The joint strategy of a subset T of $A \times \mathcal{R}$ (for instance of agents in $N(i)$ playing their roles in $SocialContext(i)$), is a tuple of local strategies, one for each agent playing a role in that set, denoted by c^t_T (e.g. $c^t_{N(i)}$). The joint strategy for all agents A at time t is denoted c^t, while the set of all joint strategies for A is the set *Strategy*.
- A state transition function $T : State \times Strategy \times State \to [0, 1]$ gives the transition probability $p(s^{t+1}|s^t, c^t)$, based on the joint strategy c^t taken in state s^t.
- A reward function per agent-role $i{:}k$ given its decisions concerning role $R_m \in SocialContext(i)$, denoted by $Rwd_{(i:k),(\cdot:m)}$, where $i \in A$ and R_k a role played by agent i. The reward function per agent-role $i{:}k$, denoted by $Rwd_{(i:k)}$ provides the agent $i{:}k$ with an individual reward based on the joint decision of its neighborhood, taken in its local state. The local reward of an agent i, Rwd_i, is the sum of its rewards for all the roles it plays.

It must be noticed that states represent agents' assumptions about periods' correspondences, while agents' strategies concern the specific periods for scheduling role-specific tasks. The reward function concerns decisions made by agents, i.e. agents' strategies w.r.t. their states, and depends on the utility of agents' choices while playing specific roles, on the feedback received from neighbors, and on the payoff received after performing the scheduled tasks:

$Rwd_{(i:k)}(P, s_i) = a * U(i:k, P) + b * Feedback(i:k, s_i) + Payoff(i:k, c_i)$, where $Feedback(i:k, s_i) = Payoff * Feedback^+(i:k, s_i) + Penalty * Feedback^-(i:k, s_i)$, $P \in \mathcal{P}_{R_i}$, and $Feedback^+(i:k, s_i)$, $Feedback^-(i:k, s_i)$ are the numbers of positive and negative feedbacks received, respectively, $Payoff$ and $Penalty$ are the numbers specifying the payoff and cost for each positive and negative feedback, respectively (being equal to the corresponding utility parameters). The parameters a and b have been used for balancing between own utility and feedback received by others: As previous works have shown [5], the role of both is crucial. The method is tolerant to different values of these parameters, but here we consider that $\frac{a}{b} = \frac{1}{10}$. Finally, $Payoff(i:k, c_i)$ is the payoff that the agent i receives after performing R_k tasks by applying the strategies chosen.

A (local) policy of an agent i in its social context is a function $\pi_i : s_i \to c_i$ that returns a local decision for any given local state. The objective for any agent in the society is to find an optimal policy π^* that maximizes the expected discounted future return $V_i^*(s) = max_{\pi_i} E[\sum_{t=0}^{\infty} \delta^t Rwd_i(\pi_i(s_i^t), s_i^t)|\pi_i)]$ for each state s_i, while playing all its roles. The expectation $E(.)$ averages over stochastic transitions, and $\delta \in [0, 1]$ is the discount factor.

This model assumes the Markov property, assuming also that rewards and transition probabilities are independent of time. Thus, the state next to state s is denoted by s' and it is independent of time.

Q-functions, or action-value functions, represent the future discounted reward for a state s when making the choice c and behaving optimally from then on. The optimal policy for the agents in state s is to jointly make the choice $argmax_c Q^*(s, c)$ that maximizes the expected future discounted reward. The next paragraphs describe two distributed variants of Q-learning considering that agents do not know the transition and reward model (model-free methods) and interact with their neighbors, only. Both variants assume that agents propagate their decisions to neighbors, and take advantage of dependencies with others, specified by means of the edges connecting them in the society.

Independent Reinforcement Learners: In the first variant, the local function Q_i for an agent i is defined as a linear combination of all contributions from its social context, for any role R_k played by i in combination with roles R_m in its social context: $Q_i = \sum_{R_k} \sum_{j:m, j \in N(i)} Q_{(i:k),(j:m)}$. To simplify the formulae we denote $((i:k), (j:m))$ by $i \bowtie j$. Thus, each $Q_{i \bowtie j}$ is updated as follows:

$Q_{i \bowtie j}(s_i, c_{i \bowtie j}) = Q_{i \bowtie j}(s_i, c_{i \bowtie j}) +$
$\alpha[Rwd_{i:k}(s_i, c_{i \bowtie j}) + \delta max_{c'_{i \bowtie j}} Q_{i \bowtie j}(s'_i, c_{i \bowtie j}) - Q_{i \bowtie j}(s_i, c_{i \bowtie j}))]$

This method is in contrast to the Coordinated Reinforcement Learning model proposed by Guestrin in [8] that considers society's global state, and it is closer to the model of independent learners, since the formula considers the local states of agents.

Collaborative Reinforcement Learners: The second variant is the agent-based update sparse cooperative edge-based Q-learning method proposed in [9]. Given two neighbor agents $i{:}k$ and $j{:}m$, the Q-function is denoted $Q_{i:k,j:m}(s_{i:k,j:m}, c_{i:k,j:m}, c_{j:m,i:k})$, or succinctly $Q_{i\bowtie j}(s_{i\bowtie j}, c_{i\bowtie j}, c_{j\bowtie i})$, where $s_{i\bowtie j}$ are the state variables related to the two agents playing their roles, and $c_{i\bowtie j}, c_{j\bowtie i}$ are the strategies chosen by the two agents. The sum of all these edge-specific Q-functions defines the global Q-function. It must be noticed that it may hold that $i = j$, considering the Q-functions for the different roles the agent i is playing. The update function is as follows:

$$Q_{i\bowtie j}(s_{i\bowtie j}, c_{i\bowtie j}, c_{j\bowtie i})) = Q_{i\bowtie j}(s_{i\bowtie j}, c_{i\bowtie j}, c_{j\bowtie i})) +$$
$$\alpha \sum_{x:y \in \{i:k,j:m\}} \frac{Rwd_{x:y}(s_{x:y}, c_{x:y}) + \delta Q^*_{x:y}(s'_{x:y}, c_{x:y}) - Q_{x:y}(s_{x:y}, c_{x:y})}{|N(x)|}$$

The local function of an agent $i{:}k$ is defined to be the summation of half the value of all local functions $Q_{i\bowtie j}(s_{i\bowtie j}, c_{i\bowtie j}, c_{j\bowtie i})$ for any $j{:}m$, with $j \in N(i)$ and $R_m \in SocialContext(i)$: $Q_{i:k}(s_{i:k}, c_{i:k}) = \frac{1}{2} \sum_{j:m} Q_{i\bowtie j}(s_{i\bowtie j}, c_{i\bowtie j}, c_{j\bowtie i})$. Closing this section we need to answer whether agents in any society do learn social norms via agreements in their social context: The answer is negative in case there are socially-isolated agents playing the same role. These are agents whose social context is limited to a single role. Thus, they do not interact "heavily" with the society and are somehow isolated in the neighborhoods of others. These are for instance the agents a and b in Fig. 1: They interact only with the agents $i{:}m$ and $j{:}m$. Although $i{:}m$ and $j{:}m$ may reach agreements on their role-specific strategies via the path(s) connecting them, and each one of them may reach agreements with $a{:}k$ and $b{:}k$ in their social contexts, respectively, there may not be an agreement between a and b, and thus a norm may not emerge for R_m-playing agents. Nevertheless, these agents do have separate concerns and have reached agreements in their contexts. Such cases do not exist in the experimental cases considered in the section that follows.

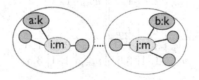

Fig. 1. Isolated agents playing the role R_k.

5   Experimental Results

We have performed simulations using the two social learning methods proposed in two types of networks: Small-world networks that have been constructed using the Watts-Strogatz model (W) [10], and scale-free networks constructed using the Albert-Barabási (B) model [11]. For both types of networks we have experimented with different populations of agents, and with various degrees of agents' connectivity. For these types of networks, we have run experiments with populations of 10, 20, 50, 100 and 200 agents, and with and average number of neighbors (ANN) 4, 10, 16, 20. Each case is denoted by $X\_|N|\_ANN$, (e.g. B_100_10) where X the network construction model. This article reports on results with B networks with different $|N|$ and $ANN = 4$, on results with B networks with $|N| = 100$ and different ANNs, and finally on $W^{0.5}$ networks with $|N| = 100$ and different ANNs.

The society roles $\mathcal{R}$ are 4, $\mathcal{R} = \{fmember, worker, dependent, boss\}$ and each agent can play up to 2 roles satisfying the following constraints: Any *worker* can be an *fmember* and vise-versa, a *dependent* can not play any other role, while a *boss* cannot play other roles and is connected to agents playing the role of *worker*. The *dependents* are up to 10 % of the population. Using these constraints, roles are assigned to agents randomly. The incompatible pairs of roles are $(fmember, worker)$, $(fmember, boss)$, $(dependent, worker)$, $(dependent, boss)$. Thus, any agent connected to an *fmember* and a *boss*, for instance, can not schedule tasks for these two roles during the same period.

We do provide results when all agents are Independent Reinforcement-Learners (IRL) or Collaborative Reinforcement-Learners (CRL). In both methods the payoff *Payoff* for positive feedback and satisfaction of constraints is equal to 3, while the penalty *Penalty* is equal to -5. Considering the reward, as already said, the ratio between the utility factor a and the feedback factor b is 1:10. For each role there are two distinct periods: The preferred (p) and the non preferred (np). These are denoted by the initial role of the role and a subscript p or np. For instance w_p is the workers preferred period. The joint task that agents need to perform is scheduling their meetings. The payoff matrices for role-specific strategies are given below.

	b_p	b_{np}			m_p	m_{np}
w_p	-1,-1	3,2		w_p	2,3	-1,-1
w_{np}	2,3	-1,-1		w_{np}	-1,-1	3,2

	d_p	d_{np}			d_p	d_{np}
w_p	3,3	-1,-1		m_p	3,3	-1,-1
w_{np}	-1,-1	3,3		m_{np}	-1,-1	3,3

It must be noticed that agents play different types of games while interacting with other roles, and do not exploit the payoffs of others in their neighborhood. Both learning methods use an exploration function, counting the number of times each correspondence or strategy has been used. An epoch comprises an exploration followed by a pure exploitation period, while the number of times that correspondences and strategies are to be tried increases by a constant in each epoch.

Figure 2 shows the results of both methods for different types of networks and different percentages of converging agents (T): The first (second) column reports on methods convergence when $T = 100\%$ (respectively, when $T = 90\%$). It must be noticed that results concerning state of the art methods require that $T \leq 90\%$. The convergence rule is that the required percentage of agents has reached agreement without violating any constraint in 10 subsequent rounds during an exploitation period. Each point in any line is the average total payoff in 5 independent runs per case received by the agents at the end of an epoch. The reported results concern 9 epochs (1000 rounds), aiming to show the efficacy of the proposed methods. A line in Fig. 2 stops at an epoch (notice that in some cases the X-axis has less than 9 points), when the corresponding method has converged in all independent runs for the corresponding case until this epoch. The average convergence round per case and method are reported in Fig. 3. The value 1000 means that the corresponding method has not managed to converge until epoch 9 (the 1000th round).

Experimentation results show that both methods are very effective both in agents convergence rate (i.e. percentage of agents reaching agreement) and in the number of epochs required. All cases converge, and in case we require 90% convergence, agents using any of the methods managed to converge to agreements in fewer than 9 epochs, except in networks with low ANN. Specifically, regarding the B networks with different populations (first two rows), as it is expected, the convergence is slower as the population increases. For networks of 100 agents, with a varying ANN, IRL converges faster for networks with higher ANN, while CRL is not affected by the degree of agents connectivity, although it converges slower than IRL in most cases when $T = 90\%$. For W networks, both methods converge less effectively. However, CRL manages to convergence more effectively when 90% convergence is required, although this is not always the case: We can observe that in networks with a large population of agents and with high ANN, IRL can be more efficient. This is reported in all cases (especially for $T = 90\%$) for B networks, but not for W networks.

6   Related Work

To frame the existing computational models towards the emergence of norms, as pointed out in [12], these may be categorized to imitation, normative advise, machine learning and data-mining models. In this paper we propose social reinforcement learning approaches to computing norms, where agents learn collaboratively by interacting in their social contexts.

Early approaches towards learning norms either involve two agents iteratively playing a stage game towards reaching a preferred equilibrium, or models where the reward of each individual agent depends on the joint action of *all* the other agents in the population. Other approaches consider that agents learn by iteratively interacting with a single opponent from the population [3], also considering the distance between agents [2]. In contrast to this, in [4] the communication between agents is physically constrained and agents interact and learn with all

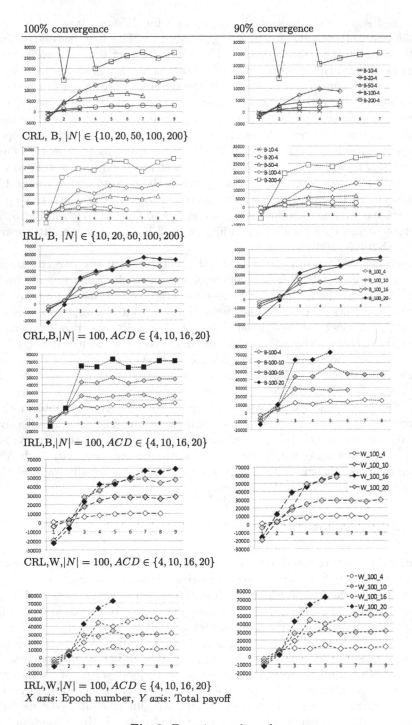

Fig. 2. Experimental results.

	CRL		IRL	
	100%	90%	100%	90%
B_10_4	278.3	229.8	287.8	287
B_20_4	518.8	544	711.5	527
B_50_4	680	466.7	948.3	576
B_100_4	900.6	534	1000	660
B_200_4	1000	769.4	1000	738

	CRL		IRL	
	100%	90%	100%	90%
B_100_4	900.6	534	1000	660
B_100_10	902.6	600.4	950.8	527
B_100_16	710.2	627.8	750	505
B_100_20	958.8	718.4	725.4	372

Fig. 3. Average convergence round per case.

their neighbors. In these works agents learn rules of the road by playing a single role at each time step. We rather consider more realistic cases where agents do not share knowledge of their environment, they play multiple roles and interact with all their neighbors who also play multiple and maybe incompatible roles simultaneously. Finally, agents have role-specific preferences on their strategies.

Concerning the learning methods that have been used, Shoham and Tennenholtz [13] proposed a reinforcement learning approach using the Highest Cumulative Reward rule. However this rule depends on the memory size of agents, as far as the history of agents' past strategy choices is concerned. The effects of memory and history of agents' past actions have also been considered in the work reported by Villatoro et al. [14,15]. Sen et al. [3] studied the effectiveness of reinforcement methods also considering the influence of the population size, of the possible actions, the existence of different types of learners in the population, as well as the underlying network topology of agents [16]. In [4] authors have proposed a learning method where each agent, at each time step interacts with all its neighbors simultaneously and use ensemble learning methods to compute a final strategy.

These studies (e.g. [2–4]), have shown that Q-learners are more efficient than other learners using for instance WoLF [17], Fictitious Play [18], Highest Cumulative Reward-based [13] models. Based on these conclusions and going beyond the state of the art, this work proposes two social Q-learning methods, according to which agents interact with all of their neighbors, considering their roles in their social contexts. Agents compute role-specific strategies, while for a single role the decisions taken depend on the feedback received from others, the existing constraints and role-specific preferences. To further advance the state of the art and study the emergence of conventions in open societies where agents do not share common representations of the world, we incorporate the computation of semantic agreements towards learning effective conventions.

7 Conclusions and Further Work

This article proposes two social, distributed reinforcement learning methods for agents to compute conventions concerning the use of common resources to perform joint tasks. The computation of agreed conventions is done via reaching agreements in agents' social context, via interactions with acquaintances playing their roles. The formulated methods support agents to play multiple roles simultaneously; even roles with incompatible requirements and different preferences on the use of resources. In conjunction to the above, and to a greater extent than state of art models, the article considers open agent societies where agents do not share common representations of the world: This necessitates the computation of semantic agreements (i.e. agreements on the meaning of terms representing resources), which is addressed with the computation of emergent conventions in an intertwined manner. Experimental results show the efficiency of both social learning methods, even if we require all agents in the society to reach agreements, despite the complexity of the problem considered. Indeed, the proposed methods require few epochs, even when we require 100 % convergence, w.r.t. the number of agents in the society. However the effectiveness of convergence is affected by both, the structure of the network and the average number of neighbors (ANN) per agent. An interesting remark is that in networks with a large population of agents and with high ANN (i.e. in highly constrained settings), methods may be more effective (this is more clear for individual learners in scale-free networks). Further experimentation is necessary to reach conclusive results regarding the specific problem parameters that affect methods effectiveness.

Further work concerns investigating (a) the effectiveness of hierarchical reinforcement learning techniques [19] for computing hierarchical policies (for correspondences, scheduling strategies and joined tasks); (b) the tolerance of the methods to different payoffs of performing joined tasks, as well as to different exploration-exploitation schemes, and (c) societies with different types of learners.

Acknowledgement. The publication of this article has been partially supported by the University of Piraeus Research Center.

References

1. Epstein, J.: Learning to be thoughtless: social norms and individual computation. Comput. Econ. **18**(1), 9–24 (2001)
2. Mukherjee, P., Sen, S., Airiau, S.: Norm emergence under constrained interactions in diverse societies. In: Padgham, L., Parkes, D.C., Müller, J.P., Parsons, S. (eds.) AAMAS (2), pp. 779–786. IFAAMAS (2008)
3. Sen, S., Airiau, S.: Emergence of norms through social learning. In: Proceedings of the 20th International Joint Conference on Artifical Intelligence, IJCAI 2007, pp. 1507–1512. Morgan Kaufmann Publishers Inc., San Francisco (2007)
4. Yu, C., Zhang, M., Ren, F., Luo, X.: Emergence of social norms through collective learning in networked agent societies. In: Proceedings of the 2013 International Conference on Autonomous Agents and Multi-agent Systems, AAMAS 2013, pp. 475–482. International Foundation for Autonomous Agents and Multiagent Systems, Richland (2013)
5. Vouros, G.: Decentralized semantic coordination via belief propagation. In: Proceedings of the 2013 International Conference on Autonomous Agents and Multi-agent Systems, AAMAS 2013, pp. 1207–1208. International Foundation for Autonomous Agents and Multiagent Systems, Richland (2013)
6. Puterman, M.L.: Markov Decision Processes: Discrete Stochastic Dynamic Programming, 1st edn. Wiley, New York (1994)

7. Guestrin, C.E.: Planning under uncertainty in complex structured environments. Ph.D. thesis, Stanford, CA, USA (2003) AAI3104233
8. Guestrin, C.G., Lagoudakis, M., Parr, R.: Coordinated reinforcement learning. In: Proceedings of the ICML-2002 The Nineteenth International Conference on Machine Learning, pp. 227–234 (2002)
9. Kok, J.R., Vlassis, N.: Collaborative multiagent reinforcement learning by payoff propagation. J. Mach. Learn. Res. **7**, 1789–1828 (2006)
10. Watts, D.J., Strogatz, S.H.: Collective dynamics of 'small-world' networks. Nature **393**(6684), 440–442 (1998)
11. Albert, R., Lászl Barabási, A.: Statistical mechanics of complex networks. Rev. Mod. Phys. **74**, 47–97 (2002)
12. Savarimuthu, B.T.R.: Norm learning in multi-agent societies. Information Science Discussion Papers Series No. 2011/05 (2011). http://hdl.handle.net/10523/1690 (retrieved)
13. Shoham, Y., Tennenholtz, M.: On the emergence of social conventions: modeling, analysis, and simulations. Artif. Intell. **94**(1–2), 139–166 (1997)
14. Villatoro, D., Sabater-Mir, J., Sen, S.: Social instruments for robust convention emergence. In: Proceedings of the Twenty-Second International Joint Conference on Artificial Intelligence, IJCAI 2011, vol. 1, pp. 420–425. AAAI Press (2011)
15. Villatoro, D., Sen, S., Sabater-Mir, J.: Topology and memory effect on convention emergence. In: Proceedings of the 2009 IEEE/WIC/ACM International Joint Conference on Web Intelligence and Intelligent Agent Technology, WI-IAT 2009, vol. 02, pp. 233–240. IEEE Computer Society, Washington, DC (2009)
16. Sen, O., Sen, S.: Effects of social network topology and options on norm emergence. In: Padget, J., Artikis, A., Vasconcelos, W., Stathis, K., da Silva, V.T., Matson, E., Polleres, A. (eds.) COIN@AAMAS 2009. LNCS, vol. 6069, pp. 211–222. Springer, Heidelberg (2010)
17. Bowling, M., Veloso, M.: Multiagent learning using a variable learning rate. Artif. Intell. **136**, 215–250 (2002)
18. Fudenberg, D., Levine, D.: The Theory in Learning in Games. The MIT Press, Cambridge (1998)
19. Barto, A.G., Mahadevan, S.: Recent advances in hierarchical reinforcement learning. Discrete Event Dyn. Syst. **13**(1–2), 41–77 (2003)

Author Index

Printed in the United States
By Bookmasters